Lecture Notes in Computer Science ⌐813

Commenced Publication in 1973
Founding and Former Series Editors:
Gerhard Goos, Juris Hartmanis, and Jan

Editorial Board

David Hutchison
Lancaster University, UK

Takeo Kanade
Carnegie Mellon University, Pittsburgh, PA, USA

Josef Kittler
University of Surrey, Guildford, UK

Jon M. Kleinberg
Cornell University, Ithaca, NY, USA

Friedemann Mattern
ETH Zurich, Switzerland

John C. Mitchell
Stanford University, CA, USA

Moni Naor
Weizmann Institute of Science, Rehovot, Israel

Oscar Nierstrasz
University of Bern, Switzerland

C. Pandu Rangan
Indian Institute of Technology, Madras, India

Bernhard Steffen
University of Dortmund, Germany

Madhu Sudan
Massachusetts Institute of Technology, MA, USA

Demetri Terzopoulos
New York University, NY, USA

Doug Tygar
University of California, Berkeley, CA, USA

Moshe Y. Vardi
Rice University, Houston, TX, USA

Gerhard Weikum
Max-Planck Institute of Computer Science, Saarbruecken, Germany

Refik Molva   Gene Tsudik
Dirk Westhoff (Eds.)

Security and Privacy in Ad-hoc and Sensor Networks

Second European Workshop, ESAS 2005
Visegrad, Hungary, July 13-14, 2005
Revised Selected Papers

 Springer

Volume Editors

Refik Molva
Institut Eurécom
2229 Route des Crêtes
06560 Valbonne Sophia Antipolis, France
E-mail: molva@eurecom.fr

Gene Tsudik
University of California, Irvine
Computer Science Department
Irvine CA 92697-3425, USA
E-mail: gts@ics.uci.edu

Dirk Westhoff
NEC Europe Ltd., Network Laboratories
Kurfürsten-Anlage 36
69115 Heidelberg, Germany
E-mail: dirk.westhoff@netlab.nec.de

Library of Congress Control Number: 2005937512

CR Subject Classification (1998): E.3, C.2, F.2, H.4, D.4.6, K.6.5

ISSN   0302-9743
ISBN-10   3-540-30912-8 Springer Berlin Heidelberg New York
ISBN-13   978-3-540-30912-3 Springer Berlin Heidelberg New York

This work is subject to copyright. All rights are reserved, whether the whole or part of the material is
concerned, specifically the rights of translation, reprinting, re-use of illustrations, recitation, broadcasting,
reproduction on microfilms or in any other way, and storage in data banks. Duplication of this publication
or parts thereof is permitted only under the provisions of the German Copyright Law of September 9, 1965,
in its current version, and permission for use must always be obtained from Springer. Violations are liable
to prosecution under the German Copyright Law.

Springer is a part of Springer Science+Business Media

springer.com

© Springer-Verlag Berlin Heidelberg 2005
Printed in Germany

Typesetting: Camera-ready by author, data conversion by Scientific Publishing Services, Chennai, India
Printed on acid-free paper   SPIN: 11601494   06/3142   5 4 3 2 1 0

Preface

It was a pleasure to take part in the 2005 European Workshop on Security and Privacy in Ad Hoc and Sensor Networks (ESAS 2005), held on July 13–14 in Visegrad (Hungary) in conjunction with the First International Conference on Wireless Internet (WICON) <http://www.wicon.org/>.

As Program Co-chairs, we are very happy with the outcome of this year's ESAS workshop. It clearly demonstrates the continued importance, popularity and timeliness of the workshop's topic: security and privacy in ad hoc and sensor networks. A total of 51 full papers were submitted. Each submission was reviewed by at least three expert referees. After a short period of intense discussions and deliberations, the Program Committee selected 17 papers for presentation and subsequent publication in the workshop proceedings. This corresponds to an acceptance rate of 33% — a respectable rate by any measure.

First and foremost, we thank the authors of ALL submitted papers. Your confidence in this venue is much appreciated. We hope that you will continue patronizing ESAS as authors and attendees. We are also very grateful to our colleagues in the research community who served on the ESAS Program Committee. Your selfless dedication is what makes the workshop a success.

Finally, we are very grateful to the ESAS Steering Group: Levente Buttyan, Claude Castelluccia, Dirk Westhoff and Susanne Wetzel. They had the vision and the drive to create this workshop in the first place; they also provided many insights and lots of help with this year's event. We especially acknowledge and appreciate the work of Levente Buttyan whose dedication (as Steering Committee member, PC member and Local Arrangements Chair) played a very important role in the success of the workshop.

September 2005

Refik Molva
Gene Tsudik

Organization

Program Chairs

Refik Molva, Eurecom, France
Gene Tsudik, UC Irvine, USA

Program Committee

Imad Aad, EPFL, Switzerland
N. Asokan, Nokia, Finland
Sonja Buchegger, UC Berkeley, USA
Laurent Bussard, Microsoft, Germany
Levente Buttyán, BUTE, CrySyS Lab, Hungary
Srdjan Capkun, UCLA, USA
Claude Castelluccia, INRIA, France
Hannes Hartenstein, University of Karlsruhe, Germany
Yih-Chun Hu, UC Berkeley, USA
Markus Jakobsson, Indiana University, Bloomington, USA
Yongdae Kim, University of Minnesota, Minneapolis, USA
Stefan Lucks, University of Mannheim, Germany
Breno de Medeiros, Florida State University, USA
Ludovic M, Supelec, France
Gabriel Montenegro, SunLabs, USA
Cristina Nita-Rotaru, Purdue University, USA
Guevara Noubir, Northeastern University, USA
Kaisa Nyberg, Nokia, Finland
Christof Paar, University of Bochum, Germany
Panagiotis Papadimitratos, Cornell University, USA
Andre Weimerskirch, University of Bochum, Germany
Dirk Westhoff, NEC Europe Network Lab., Germany
Susanne Wetzel, Stevens Institute of Technology, USA

Workshop Organizers

Levente Buttyán, Budapest University of Technology and Economics, Hungary
(buttyan@crysys.hu)

Claude Castelluccia, INRIA, France (Claude.Castelluccia@inrialpes.fr)

Dirk Westhoff, NEC Europe Network Lab., Heidelberg, Germany
(Dirk.Westhoff@netlab.nec.de)

Susanne Wetzel, Stevens Institute of Technology, USA (swetzel@cs.stevens.edu)

Table of Contents

Efficient Verifiable Ring Encryption for Ad Hoc Groups

Joseph K. Liu[1], Patrick P. Tsang[1], and Duncan S. Wong[2]

[1] Department of Information Engineering,
The Chinese University of Hong Kong Shatin, Hong Kong
{ksliu, pktsang3}@ie.cuhk.edu.hk
[2] Department of Computer Science,
City University of Hong Kong, Kowloon, Hong Kong
duncan@cityu.edu.hk

Abstract. We propose an efficient *Verifiable Ring Encryption* (VRE) for ad hoc groups. VRE is a kind of verifiable encryption [16,1,4,2,8] in which it can be publicly verified that there exists at least one user, out of a designated group of n users, who can decrypt the encrypted message, while the semantic security of the message and the anonymity of the actual decryptor can be maintained. This concept was first proposed in [10] in the name of *Custodian-Hiding Verifiable Encryption*. However, their construction requires the inefficient cut-and-choose methodology which is impractical when implemented. We are the first to propose an efficient VRE scheme that does not require the cut-and-choose methodology.

In addition, while [10] requires interaction with the encryptor when a verifier verifies a ciphertext, our scheme is non-interactive in the following sense: (1) an encryptor does not need to communicate with the users in order to generate a ciphertext together with its validity proof; and (2) anyone (who has the public keys of all users) can verify the ciphertext, without the help of the encryptor or any users. This non-interactiveness makes our scheme particularly suitable for ad hoc networks in which nodes come and go frequently as ciphertexts can be still generated and/or verified even if other parties are not online in the course. Our scheme is also proven secure in the random oracle model.

1 Introduction

A *Verifiable Encryption* [16,1,4,2,8] allows a prover to encrypt a message and sends to a receiver such that the ciphertext is publicly verifiable. That is, any verifier can ensure the ciphertext can be decrypted by the receiver yet knowing nothing about the plaintext. There are numerous applications of verifiable encryption. For example, in a publicly verifiable secret sharing scheme [16], a dealer shares a secret with several parties such that a third party can verify that the sharing was done correctly. This can be done by verifiably encrypting each shares under the public key of the corresponding party and proves to the third party that the ciphertext encrypt the correct shares. Another scenario is in a fair exchange environment [1], in which both parties want to exchange some

R. Molva, G. Tsudik, and D. Westhoff (Eds.): ESAS 2005, LNCS 3813, pp. 1–13, 2005.
© Springer-Verlag Berlin Heidelberg 2005

information such that either each party obtain the other's data, or neither party does. One approach is to let both parties verifiably encrypt their data to each other under the public key of a trusted party and then to reveal their data. If one party refuses to do so, the other can go to the trusted party to obtain the required data. Verifiable encryption can be also applied in revokable anonymous credential [5]. When the administration organization issues a credential, it verifiably encrypts enough information under the public key of the anonymity revocation manager, so that later if the identity of the credential owner needs to be revealed, this information can be decrypted.

In an interactive *Custodian-Hiding Verifiable Encryption* (CHVE) [10], an *Encryptor* wants to send a public-key encrypted message to one among a group of n *users* through a *Verifier*. The Encryptor plays the role of a *Prover* and conducts an interactive protocol with the Verifier such that, if the Verifier is satisfied, at least one of the n possible decryptors can recover the message. At the same time, the message is semantically secure, even against the Verifier, and the identity of the actual decryptor is anonymous, again even to the Verifier. Custodian-Hiding Verifiable Encryption can be found useful in the applications of gateway system or receiver-oblivious transfer.

In ad hoc networks, nodes are highly dynamic and may switch from being online and being offline frequently from time to time. The verifiability of interactive Custodian-Hiding Verifiable Encryption schemes is virtually of no practical use if the encryptor goes offline (or leaves the networks forever) since no one can verify the validity of the ciphertext without the help of the encryptor. In the environment of ad hoc networks in which most users are highly mobile, it is unreasonable to require an encryptor to be always online and available to be contacted by a verifier. What we need is exactly a non-interactive approach to verify the ciphertext.

Let us spare a few words explaining the decision of naming our scheme as "Verifiable Ring Encryption" over "Custodian-Hiding Verifiable Encryption", as suggested by [10]. The word "Ring" is borrowed from Ring signatures [15] which is a signature scheme constructed in the structure of a ring in order to achieve 1-out-of-n anonymity of the signer. Analogously, Verifiable Ring Encryption implies an encryption scheme constructed in the structure of a ring, in which ciphertexts can be verified to be decryptable by some one, with the identity of that genuine decryptor hidden among a group of n members. Our choice of "Verifiable Ring Encryption" therefore better conveys the information on what the scheme actually does. Moreover, the non-interactiveness of our scheme suggests that a verifier is convinced by verifying the validity of some kind of proofs. These proofs can actually be thought of a kind of ring signatures in the sense that they convince verifiers of the fact that some 1 out of n users can decrypt a ciphertext, and yet hiding that decryptor's identity.

Finally we would like to note that the notion of "Verifiable Group Encryption" (VGE) has been used by [4] to mean something related but very different: VGE allows the prover to prove that any subset of t members of a group of n users can jointly recover the message behind a ciphertext, by making use of a secret sharing scheme. That is, the prover divides the message into n pieces of shares such that any t of them are enough to reconstruct m. Then he encrypts each share for each user using the user's encryption function, and sends all ciphertexts to the verifier. It is clear that the message m can be reconstructed if any t users decrypt their corresponding ciphertext to get the shares.

1.1   Contributions

We propose an efficient Verifiable Ring Encryption for ad hoc networks which is the first of its kind that is without the use of the inefficient cut-and-choose methodology. Furthermore, our proposed scheme is non-interactive. Unlike the previous one proposed in [10], in our scheme an encryptor does not need to communicate with the users in order to generate a ciphertext together with its validity proof. Also anyone who has got the public keys of all users can verify the ciphertext without the help of the encryptor or any users. Note that being non-interactive makes our scheme well-suited for ad hoc networks in which nodes are highly mobile. Ciphertexts can be still generated and/or verified even if other parties are not online in the course. We also prove the security of our proposed scheme in the random oracle model [3].

Organization: The rest of the paper is organized as follows. We give security definitions in Sec. 2. The details of our proposed scheme is presented in Sec. 3. Its security is analyzed in Sec. 4. We conclude the paper in Sec. 5.

2   Security Definition

2.1   Notations

Let a be a real number. We denote by $\lfloor a \rfloor$ the largest integer $b \leq a$, by $\lceil a \rceil$ the smallest integer $b \geq a$, and by $\lfloor a \rceil$ the largest integer $b \leq a + 1/2$. For positive real numbers a and b, let $[a]$ denote the set $\{0, 1 \ldots, \lfloor a \rfloor - 1\}$ and $[a, b]$ the set $\{\lfloor a \rfloor, \ldots, \lfloor b \rfloor\}$ and $[-a, b]$ denote the set $\{-\lfloor a \rfloor, \ldots, \lfloor b \rfloor\}$.

By $\text{neg}(\lambda)$ we denote a negligible function, i.e., a function f such that $f(\lambda) < 1/p(\lambda)$ holds for all polynomials $p(\lambda)$ and all sufficiently large λ.

We also use the shorthand notation $\{PK\}_N$ and $\{SK\}_N$, $N \in \mathbb{N}$, to mean the sets $\{PK_1, \ldots, PK_N\}$ and $\{SK_1, \ldots, SK_N\}$ respectively.

2.2   A High Level Description

Before giving a formal definition of verifiable ring encryption, we begin with a high level discussion of this notion in order to let readers understand more easily.

We start by the description of an ordinary verifiable encryption. A verifiable encryption scheme proves that a ciphertext encrypts a plaintext satisfying a certain relation $\mathcal{R}$. The relation $\mathcal{R}$ is defined by a generator algorithm $\mathcal{G}'$ which on input a security parameter λ outputs a binary relation $W \times \Delta$. For $\delta \in \Delta$, an element $w \in W$ such that $(w, \delta) \in \mathcal{R}$ is called a *witness* for δ. The encryptor will be given a value δ, a witness w for δ, then encrypts w to generate a ciphertext

ψ. Later, the encryptor may prove to another party that ψ decrypts to a witness for δ. In this system, the honest verifier will output accept or reject. If the system is sound, the verifier accepts a proof means that with overwhelming probability the ciphertext ψ can be decrypted to a witness for δ.

We extend this concept into a group of N designated receivers. In a verifiable ring encryption scheme, a prover proves that a ciphertext encrypts a plaintext satisfying one of the certain relation $\mathcal{R}$ which is corresponding to one of the receiver. The idea is that the encryptor will be given a value w, which is a witness for δ where $(w, \delta) \in \mathcal{R}$, and randomly generates other $N-1$ witnesses and the corresponding group elements.

Note that for an interactive proof system, both the prover and the verifier are required to interact in order to have the verifier convinced. If the proof system is non-interactive, the proof is carried out in a non-interactive fashion – the prover (or the encryptor) generates a proof transcript that can be used to convince a verifier at any later time that one (out of N) of the receivers can decrypt the corresponding witness of that group element δ. However, the verifier still cannot compute the identity of the actual decryptor.

2.3   Defining Verifiable Ring Encryption

A *Verifiable Ring Encryption* (VRE) scheme is actually a group encryption scheme with add-on *Verifiability*. A group encryption scheme is a generalization of a public key encryption scheme. Entities involved in such a scheme include an *encryptor* and a group of N *users*. The encryptor has a secret message m which he wants to send to a certain designated one out of the N users in the group, so that the secret message can be decrypted only by the designated member. In other words, a VRE scheme, apart from allowing a secret message to be encrypted to some designated members, provides with the encryptor the ability to prove that a ciphertext encrypts a plaintext satisfying certain relation $\mathcal{R}$.

The relation $\mathcal{R}$ is defined by a generator algorithm $\mathcal{G}'$ which on input 1^λ outputs a description $\Psi = \Psi[\mathcal{R}, W, \Delta]$ of a binary relation $\mathcal{R}$ on $W \times \Delta$. We require that the sets $\mathcal{R}$, W, and Δ are easy to recognize (given Ψ). For $\delta \in \Delta$, an element $w \in W$ such that $(w, \delta) \in \mathcal{R}$ is called a witness for δ. The idea is that the encryptor will be given a value δ, a witness w for δ, and a label L, and then encrypts w under L, yielding a ciphertext ψ. After this, the encryptor may prove to another party that ψ decrypts under L to a witness for δ. In carrying out the proof, the encryptor will need to make use of the random coins that were used by the encryption algorithm.

Now, a Ver-Gp-Enc scheme is a tuple of $(\mathcal{S}, \mathcal{G}, \mathcal{E}, \mathcal{D}, \mathcal{P}, \mathcal{V})$ defined as follows:

- param $\leftarrow \mathcal{S}(1^\lambda)$, the probabilistic polytime (PPT) *Setup* algorithm that on input security parameter 1^λ, $\lambda \in \mathbb{N}$, outputs and publishes a set of system's parameters param that also includes the security parameter 1^λ, and a description $\Psi[\mathcal{R}, W, \Delta] \leftarrow \mathcal{G}'(1^\lambda)$.
- $(\mathsf{PK}_i, \mathsf{SK}_i) \leftarrow \mathcal{G}(\mathsf{param}, 1^{\lambda_i})$, the PPT *Key Generation* algorithm that on input the set of system's parameters param and security parameter 1^{λ_i}, $\lambda_i \in \mathbb{N}$,

where $\lambda_i \geq \lambda$, outputs a public-key/private key pair $(\mathsf{PK}_i, \mathsf{SK}_i)$. PK_i includes also the security parameter 1^{λ_i}.

- $\psi \leftarrow \mathcal{E}(\mathsf{param}, N, \{\mathsf{PK}\}_N, \pi, w, \delta, L)$, the PPT *Encryption* algorithm that takes as input the set of system's parameters param, the group size $N \in \mathbb{N}$ of size polynomial in λ, a set of N public keys $\{\mathsf{PK}\}_N$, an index $\pi \in [1, N]$, a message $w \in W$ which is the witness of $\delta \in \Delta$, and a label $L \in \{0, 1\}^*$, and outputs a ciphertext ψ. We denote by $\mathcal{E}'(\mathsf{param}, N, \{\mathsf{PK}\}_N, \pi, w, \delta, L)$ the pair (ψ, coins), where ψ is the output of $\mathcal{E}(\mathsf{param}, N, \{\mathsf{PK}\}_N, \pi, w, \delta, L)$ and coins are the random coins used by $\mathcal{E}$ to compute ψ.

- $m/\perp \leftarrow \mathcal{D}(\mathsf{param}, N, \{\mathsf{PK}\}_N, \pi, \mathsf{SK}_\pi, \psi, L)$, the polynomial-time *Decryption* algorithm that takes as input the set of system's parameters param, the group size $n \in \mathbb{N}$ of size polynomial in λ, a set $\{\mathsf{PK}\}_N$ of N public keys, an index $\pi \in [1, N]$, a private key SK_π, a ciphertext ψ, and a label $L \in \{0, 1\}^*$, and outputs either a message $m \in \mathcal{M}$, or a special symbol $\perp$. The output of the algorithm implicitly defines the domain of m, that we denote by $\mathcal{M}$.

- $\mathsf{proof} \leftarrow \mathcal{P}(\mathsf{param}, N, \{\mathsf{PK}\}_N, \pi, w, \delta, L, \psi, \mathit{coins})$, the PPT *Proof* algorithm that takes as input the tuple $(\mathsf{param}, N, \{\mathsf{PK}\}_N, \pi, w, \delta, L, \psi, \mathit{coins})$ such that (ψ, coins) is the output of some $\mathcal{E}'(\mathsf{param}, N, \{\mathsf{PK}\}_N, \pi, w, \delta, L)$, and outputs a proof proof.

- $0/1 \leftarrow \mathcal{V}(\mathsf{param}, N, \{\mathsf{PK}\}_N, L, \psi, \mathsf{proof})$, the polynomial-time *Verification* algorithm that takes as input the tuple $(\mathsf{param}, N, \{\mathsf{PK}\}_N, \pi, L, \psi)$ such that ψ is the output of some $\mathcal{E}(\mathsf{param}, N, \{\mathsf{PK}\}_N, \pi, w, \delta, L)$ for some $\pi \in [1, N]$, $w \in \mathcal{M}$ and $\delta \in \Delta$, and outputs either 0 or 1, indicating accept or reject respectively.

Here we take a more relaxed approach in order to make it to be more convenient and adequate for practical applications. Instead of requiring the ciphertext to be decrypted to a witness, we only require that a witness can be easily reconstructed from the plaintext using some efficient reconstruction algorithm *recon*. We believe that this definition is more suitable for many applications.

Definition 1. *The above* Ver-Gp-Enc *scheme is a Verifiable Group Encryption scheme, if it is (1) correct, (2) sound, (3) zero-knowledge and (4) anonymous, as defined in the following.*

Correctness: *A* Ver-Gp-Enc *is correct if it satisfies both* Verification Correctness *and* Decryption Correctness *defined below.*

- (Verification Correctness.) *For all* $\mathsf{param} \leftarrow \mathcal{S}(1^\lambda)$, *for all* $N \in \mathbb{N}$ *of size polynomial in* λ, *for all* $\lambda_i \geq \lambda$, $i \in [1, N]$, *for all* $(\mathsf{PK}_i, \mathsf{SK}_i) \leftarrow \mathcal{G}(\mathsf{param}, 1^{\lambda_i})$, $i \in [1, N]$, *for all* $(w, \delta) \in \mathcal{R}$, *for all* $L \in \{0, 1\}^*$, *for all* $\pi \in [1, N]$, *for all* $(\psi, \mathit{coins}) \leftarrow \mathcal{E}'(\mathsf{param}, N, \{\mathsf{PK}\}_N, \pi, w, \delta, L)$, *for all*

$$\mathsf{proof} \leftarrow \mathcal{P}(\mathsf{param}, N, \{\mathsf{PK}\}_N, \pi, w, \delta, L, \psi, \mathit{coins}),$$

$$\Pr[x \leftarrow \mathcal{V}(\mathsf{param}, N, \{\mathsf{PK}\}_N, L, \psi, \mathsf{proof}) : x = 1] = 1 - \mathrm{neg}(\lambda).$$

– (Decryption Correctness.) *For all param* $\leftarrow \mathcal{S}(1^\lambda)$, *for all* $N \in \mathbb{N}$ *of size polynomial in* λ, *for all* $\lambda_i \geq \lambda$, $i \in [1, N]$, *for all* $(PK_i, SK_i) \leftarrow \mathcal{G}(param, 1^{\lambda_i})$, $i \in [1, N]$, *for all* $\pi \in [1, N]$, *for all* $w \in \mathcal{M}$, *for all* $L \in \{0, 1\}^*$, *for all*

$$\psi \leftarrow \mathcal{E}(param, N, \{PK\}_N, \pi, w, \delta, L),$$

$$\Pr[\tilde{m} \leftarrow \mathcal{D}(param, N, \{PK\}_N, \pi, SK_\pi, \psi, L) : m = \tilde{m}] = 1 - \text{neg}(\lambda).$$

Soundness: *For all PPT adversaries* $\mathcal{A}_1, \mathcal{A}_2$, *and some reconstruction PPT algorithm recon,*

$$\Pr[\,param \leftarrow \mathcal{S}(1^\lambda);$$
$$(N, \lambda_1, \ldots, \lambda_N) \leftarrow \mathcal{A}_1(param),$$
where N *has a size polynomial in* λ *and* $\lambda_i \geq \lambda$ *for all* $i \in [1, N]$;
$$(PK_i, SK_i) \leftarrow \mathcal{G}(param, 1^{\lambda_i}), \text{ for all } i \in [1, N];$$
$$(\delta, \psi, L, proof) \leftarrow \mathcal{A}_2(param, N, \{PK\}_N, \{SK\}_N);$$
$$x \leftarrow \mathcal{V}(param, N, \{PK\}_N, L, \psi, proof);$$
$$m_j \leftarrow \mathcal{D}(param, N, \{PK\}_N, j, SK_j, \psi, L\}), \text{ for all } j \in [1, N];$$
$$w_j \leftarrow recon(param, N, \{PK\}_N, \delta, m_j), \text{ for all } j \in [1, N] :$$
$$x = 1 \wedge (\forall j \in [1, N])((w_j, \delta) \notin \mathcal{R}) \qquad\qquad]$$
$$= \text{neg}(\lambda).$$

Simply speaking, the definition of soundness above means that if a ciphertext is verified by a verifier to be valid, then there exists one user who can decrypt the ciphertext to the witness of δ, with overwhelming probability.

Zero knowledge: *There exists a PPT simulator Sim such that for all PPT adversaries* $\mathcal{A}_1, \mathcal{A}_2, \mathcal{A}_3$, *we have*

$$\Pr[\,param \leftarrow \mathcal{S}(1^\lambda);$$
$$(N, \lambda_1, \ldots, \lambda_N) \leftarrow \mathcal{A}_1(param),$$
where N *has a size polynomial in* λ *and* $\lambda_i \geq \lambda$ *for all* $i \in [1, N]$;
$$(PK_i, SK_i) \leftarrow \mathcal{G}(param, 1^{\lambda_i}), \text{ for all } i \in [1, N];$$
$$(w, \delta, L, \pi) \leftarrow \mathcal{A}_2(param, N, \{PK\}_N, \{SK\}_N),$$
where $(w, \delta) \in \mathcal{R}$, $L \in \{0, 1\}^*$ *and* $\pi \in [1, N]$;
$$(\psi, coins) \leftarrow \mathcal{E}'(param, N, \{PK\}_N, \pi, w, \delta, L);$$
$$b \leftarrow \{0, 1\};$$
if $b = 0$
    then proof $\leftarrow \mathcal{P}(param, N, \{PK\}_N, \pi, w, \delta, L, \psi, coins)$
    else proof $\leftarrow Sim(param, N, \{PK\}_N, \delta, \psi, L);$
$$\hat{b} \leftarrow \mathcal{A}_3(param, N, \{PK\}_N, \{SK\}_N, w, \delta, L, \pi, \psi, proof) :$$
$$b = \hat{b} \qquad\qquad]$$
$$= 1/2 + \text{neg}(\lambda).$$

The definition above means that an adversary cannot distinguish a simulated proof from a proof generated from real execution of algorithms. In other words, the proof is zero-knowledge to a verifier.

Anonymity: *For all PPT adversaries* $\mathcal{A}_1, \mathcal{A}_2, \mathcal{A}_3$,

$\Pr[\ param \leftarrow \mathcal{S}(1^\lambda);$

$\quad (N, \lambda_1, \ldots, \lambda_N) \leftarrow \mathcal{A}_1(param, \Psi),$

$\quad where\ N\ has\ a\ size\ polynomial\ in\ \lambda\ and\ \lambda_i \geq \lambda\ for\ all\ i \in [1, N];$

$\quad (PK_i, SK_i) \leftarrow \mathcal{G}(param, 1^{\lambda_i}),\ for\ all\ i \in [1, N];$

$\quad (w, \delta, L, \pi_0, \pi_1) \leftarrow \mathcal{A}_2(param, N, \{PK\}_N),$

$\quad where\ (w, \delta) \in \mathcal{R}\ and\ \pi_0, \pi_1 \in [1, N]\ are\ distinct;$

$\quad b \leftarrow \{0, 1\};$

$\quad (\psi, coins) \leftarrow \mathcal{E}'(param, N, \{PK\}_N, \pi_b, w, \delta, L);$

$\quad proof \leftarrow \mathcal{P}(param, N, \{PK\}_N, \pi_b, w, L, \psi, coins);$

$\quad \hat{b} \leftarrow \mathcal{A}_3(param, N, \{PK\}_N, w, \delta, L, \pi_0, \pi_1, \{SK_i | i \in [1, n] \backslash \{\pi_0, \pi_1\}\}, \psi, proof);$

$\quad \hat{b} = b \qquad\qquad\qquad\qquad\qquad\qquad\qquad\qquad\qquad\qquad\qquad\qquad\qquad]$

$= 1/2 + \mathrm{neg}(\lambda).$

The definition of anonymity above means that an adversary cannot decide better than random guessing, given a ciphertext together with a corresponding proof transcript, who among the 2 possible designated members is actually designated, even he has corrupted all of the other $(N - 2)$ members.

3   The Proposed Scheme

3.1   Key Generation

For each user, select two random ℓ-bit Sophie Germain primes p' and q', with $p' \neq q'$, and compute $p = (2p' + 1), q = (2q' + 1)$ and $n = pq$, where $\ell = \ell(\lambda)$ is a security parameter which is a polynomial in λ. Choose random $x_1, x_2, x_3 \in_R$ $[n^2/4]$, choose a random $g' \in_R \mathbb{Z}_{n^2}^*$, and compute $g = (g')^{2n}$, $y_1 = g^{x_1}$, $y_2 = g^{x_2}$ and $y_3 = g^{x_3}$.

Let Γ be a cyclic group of order ρ generated by γ. We assume ρ and γ are publicly known, and that ρ is prime. Let $W = [\rho]$ and $\Delta = \Gamma$, and let $\mathcal{R} = \{(w, \delta) \in W \times \Delta : \gamma^w = \delta\}$.

Choose two other ɩ-bit primes $\mathfrak{p}', \mathfrak{q}'$ and compute $\mathfrak{p} = 2\mathfrak{p}' + 1$, $\mathfrak{q} = 2\mathfrak{q}' + 1$ and $\mathfrak{n} = \mathfrak{p}\mathfrak{q}$, and choose $\mathfrak{g}, \mathfrak{h}$ as two generators of $\mathfrak{G}_{\mathfrak{n}'} \subset \mathbb{Z}_{\mathfrak{n}}^*$, where $\mathfrak{n}' = \mathfrak{p}'\mathfrak{q}'$ and $\mathfrak{G}_{\mathfrak{n}'}$ is the subgroup of $\mathbb{Z}_{\mathfrak{n}}^*$ of order $\mathfrak{n}'$, and $ɩ = ɩ(\lambda)$ which is a polynomial in λ.

The public key of this user is $(n, g, y_1, y_2, y_3, \mathfrak{n}, \mathfrak{g}, \mathfrak{h}, h, \rho, \gamma)$ and the secret key is (x_1, x_2, x_3, p, q) where $h = (1 + n \mod n^2) \in \mathbb{Z}_{n^2}^*$. We further define $H : \{0, 1\}^* \to \{0, 1\}^k$ be a collision resistant hash function and $\mathsf{abs} : \mathbb{Z}_{n^2}^* \to \mathbb{Z}_{n^2}^*$ maps $(a \mod n^2)$, where $0 < a < n^2$, to $(n^2 - a \mod n^2)$ if $a > n^2/2$, and to $(a \mod n^2)$, otherwise.

For a list of N users, we denote PK_i, the public key of user i be $(n_i, g_i, y_{1_i}, y_{2_i}, y_{3_i}, \mathfrak{n}_i, \mathfrak{g}_i, \mathfrak{h}_i, h_i, \rho_i, \gamma_i)$ and the corresponding secret key SK_i is $(x_{1_i}, x_{2_i}, x_{2_i}, p_i, q_i)$. For simplicity, we let L denote the list of the public keys of N users.

3.2   Encryption and Ciphertext Validity Proof

The prover sends an encrypted message to one of the N receivers such that only one of them can decrypt the message. At the same time, any verifier having the public keys of those N receivers can verify that the ciphertext can be decrypted by at least one of the receivers yet does not know the identity of this targeted receiver.

We use a special kind of encryption scheme by Camenisch and Shoup [8] where the plaintext is the discrete log of a group element. Then we apply a 1-out-of-n proof-of-knowledge methodology to achieve our goal.

To encrypt a message $m \in [n_\pi]$ under L, the list of public keys of N users, the prover executes the following algorithm:

1. For $i = 1, \ldots, N, i \neq \pi$, randomly generate $m_i \in_R [n_i]$ and compute $\delta_i = \gamma_i^{m_i}$. For π, compute $\delta_\pi = \gamma_\pi^m$.

2. Randomly generate $r_\pi, s_\pi \in_R [n_\pi/4]$ and compute $u_\pi = g_\pi^{r_\pi}$, $e_\pi = y_{1_\pi}^{r_\pi} h_\pi^m$, $v_\pi = \mathsf{abs}((y_{2_\pi} y_{3_\pi}^{H(u_\pi, e_\pi)})^{r_\pi})$, $t_\pi = \mathfrak{g}_\pi^m \mathfrak{h}_\pi^{s_\pi}$

3. Randomly generate $r'_\pi \in_R [-n_\pi 2^{k+k'-2}, n_\pi 2^{k+k'-2}]$, $s'_\pi \in_R [-n_\pi 2^{k+k'-2}, n_\pi 2^{k+k'-2}]$, $m'_\pi \in_R [-\rho_\pi 2^{k+k'-2}, \rho_\pi 2^{k+k'-2}]$ and compute $u'_\pi = g_\pi^{2r'_\pi}$, $e'_\pi = y_{1_\pi}^{2r'_\pi} h_\pi^{2m'_\pi}$, $v'_\pi = (y_{2_\pi} y_{3_\pi}^{H(u_\pi, e_\pi)})^{2r'_\pi}$, $\delta'_\pi = \gamma_\pi^{m'_\pi}$, $t'_\pi = \mathfrak{g}_\pi^{m'_\pi} \mathfrak{h}_\pi^{s'_\pi}$ and $c_{\pi+1} = H(L, \delta_\pi, u'_\pi, e'_\pi, v'_\pi, \delta'_\pi, t'_\pi)$.

4. For $i = \pi + 1, \ldots, n, 1, \ldots, \pi - 1$, randomly generate $\tilde{r}_i \in_R [-n_i 2^{k+k'-2}, n_i 2^{k+k'-2}]$, $\tilde{s}_i \in_R [-n_i 2^{k+k'-2}, n_i 2^{k+k'-2}]$, $\tilde{m}_i \in_R [-\rho_i 2^{k+k'-2}, \rho_i 2^{k+k'-2}]$, $u_i, e_i, v_i \in_R \mathbb{Z}^*_{n^2}$, $t_i \in_R \mathbb{Z}^*_{n^2}$ and compute $u'_i = u_i^{2c_i} g_i^{2\tilde{r}_i}$, $e'_i = e_i^{2c_i} y_{1i}^{2\tilde{r}_i} h_i^{2\tilde{m}_i}$, $v'_i = v_i^{2c_i} (y_{2i} y_{3i}^{H(u_i, e_i)})^{2\tilde{r}_i} \delta'_i = \delta_i^{c_i} \gamma_i^{\tilde{m}_i}$, $t'_i = t_i^{c_i} \mathfrak{g}_i^{\tilde{m}_i} \mathfrak{h}_i^{\tilde{s}_i}$, $c_{i+1} = H(L, \delta_i, u'_i, e'_i, v'_i, \delta'_i, t'_i)$

5. Compute $\tilde{r}_\pi = r'_\pi - c_\pi r_\pi$, $\tilde{s}_\pi = s'_\pi - c_\pi s_\pi$, $\tilde{m}_\pi = m'_\pi - c_\pi m_\pi$ (all are computed in $\mathbb{Z}$)

6. Output the ciphertext ψ and the proof proof, where

$$\psi := ((u_1, e_1, v_1), \ldots, (u_N, e_N, v_N)), \text{ and}$$

$$\mathsf{proof} := ((\delta_1, t_1, \tilde{r}_1, \tilde{s}_1, \tilde{m}_1), \ldots, (\delta_N, t_N, \tilde{r}_N, \tilde{s}_N, \tilde{m}_N), c_1).$$

Note that we describe the *Encryption* algorithm and the *Proof* algorithm in a combined fashion to allow a neat presentation. It should also be a common practice to do both in one shot in real applications. However, they can always be done separately if desired.

3.3 Decryption

Assume user π is the actual decryptor. To decrypt a ciphertext ψ using his own secret key SK_π, user π check whether $\mathsf{abs}(v_\pi) \stackrel{?}{=} v_\pi$ and $u_\pi^{2(x_{2\pi} + H(u_\pi, e_\pi)x_{3\pi})} \stackrel{?}{=} v_\pi^2$. If this does not hold, then output reject and halt. Next, let $t_\pi = 2^{-1} \bmod n_\pi$ and compute $\bar{m} = (e_\pi/u_\pi^{x_{1\pi}})^{2t_\pi}$. If $\bar{m}$ is of the form h_π^m for some $m \in [n_\pi]$, then output m. Otherwise, output reject.

3.4   Verification

Any verifier on input L, the list of public keys of those N users, and the ciphertext ψ, can verify that ψ can be decrypted by at least one of the N users. That is, at least one user π can reconstruct the plaintext, which is the discrete log of the group element δ_π. Yet the verifier cannot compute the identity of this actual decryptor and cannot compute the plaintext.

The verification algorithm is as follows.

1. For $i = 1, \ldots, N$, compute $\hat{u}_i = u_i^{2c_i} g_i^{2\tilde{r}_i}$, $\hat{e}_i = e_i^{2c_i} y_{1_i}^{2\tilde{r}_i} h_i^{2\tilde{m}_i}$,
   $\hat{v}_i = v_i^{2c_i} (y_{2_i} y_{3_i}^{H(u_i, e_i)})^{2\tilde{r}_i}$, $\hat{\delta}_i = \delta_i^{c_i} \gamma_i^{\tilde{m}_i}$, $\hat{t}_i = t_i^{c_i} \mathfrak{g}_i^{\tilde{m}_i} \mathfrak{h}_i^{\tilde{s}_i}$ and
   $c_{i+1} = H(L, \delta_i, \hat{u}_i, \hat{e}_i, \hat{v}_i, \hat{\delta}_i, \hat{t}_i)$ if $i \neq n$
2. Check whether
$$c_1 \stackrel{?}{=} H(L, \delta_N, \hat{u}_N, \hat{e}_N, \hat{v}_N, \hat{\delta}_N, \hat{t}_N)$$

If yes, output accept. Otherwise, output reject.

4   Security Analysis

The assumptions used for proving our scheme are the following.

Assumption 1 (Strong RSA Assumption). *Given a composite modulus n and a random element $g \in \mathbb{Z}_n^*$, it is hard to compute $h \in \mathbb{Z}_n^*$ and integer $e > 1$ such that $h^e = g$.*

Assumption 2.   (Paillier Decision Composite Residuosity (DCR) Assumption [13]) *Given only n, it is hard to distinguish random elements of $\mathbb{Z}_{n^2}^*$ from random elements of the subgroup of $\mathbb{Z}_{n^2}^*$ consisting of all n-th powers of elements in $\mathbb{Z}_{n^2}^*$.*

To be complete, one needs to specify more precisely the distribution from which n is drawn. We specify that n is of the form pq, where $p = 2p' + 1, q = 2q' + 1$, and p' and q' are uniformly distributed over all ℓ-bit numbers such that p, q, p', q' are prime and $p' \neq q'$, where ℓ is the security parameter.

Theorem 1. *Under the strong RSA and DCR assumption, our proposed scheme is a Verifiable Ring Encryption scheme in the random oracle model.*

The proof can be found in Appendix A.

5   Conclusion

In this paper, we propose a *Verifiable Ring Encryption* scheme for ad hoc groups. Different from previous verifiable encryption schemes which are for one designated receiver, our proposed scheme is targeted for a group of N receivers. However, only one of them is able to decrypt the ciphertext while others cannot. Any public verifier (who has the public keys of those N users) can verify this fact yet he cannot compute the identity of this actual decryptor. We propose a concrete construction and prove its security in the random oracle model. We believe this kind of schemes will attract many applications in practice.

In addition, there are open problems left such as constructing a verifiable ring encryption scheme that supports partial or fully separability [9,6]. To build up a verifiable subgroup encryption, that is, a targeted subgroup of t members out of a group of N members are able to decrypt the message anonymously, is another interesting future research. The physical size of the ciphertext of our proposed scheme grows linearly with the number of designated receivers. It is another open problem to make the size of the ciphertext to be irrelevant to the group size.

References

1. N. Asokan, V. Shoup, and M. Waidner. Optimistic fair exchange of digital signatures. In *Proc. EUROCRYPT 98*, pages 591–606. Springer-Verlag, 1998. Lecture Notes in Computer Science No. 1403.
2. F. Bao. An efficient verifiable encryption scheme for encryption of discrete logarithms. In *Proc. Smart Card Research and Applications (CARDIS) 1998*, pages 213–220. Springer-Verlag, 2000. Lecture Notes in Computer Science No. 1820.
3. M. Bellare and P. Rogaway. Random oracles are practical: A paradigm for designing efficient protocols. In *Proc. 1st ACM Conference on Computer and Communications Security*, pages 62–73. ACM Press, 1993.
4. J. Camenisch and I. Damgård. Verifiable encryption, group encryption, and their applications to separable group signatures and signature sharing schemes. In *Proc. ASIACRYPT 2000*, pages 331–345. Springer-Verlag, 2000. Lecture Notes in Computer Science No. 1976.
5. J. Camenisch and A. Lysyanskaya. An efficient system for non-transferable anonymous credentials with optional anonymity revocations. In *Proc. EUROCRYPT 2001*, pages 93–118. Springer-Verlag, 2001. Lecture Notes in Computer Science No. 2045.
6. J. Camenisch and M. Michels. Separability and efficiency for generic group signature schemes. In *Proc. CRYPTO 99*, pages 413–430. Springer-Verlag, 1999. Lecture Notes in Computer Science No. 1666.
7. J. Camenisch and V. Shoup. Practical verifiable encryption and decryption of discrete logarithms. http://eprint.iacr.org/2002/161/, 2002.
8. J. Camenisch and V. Shoup. Practical verifiable encryption and decryption of discrete logarithms. In *Proc. CRYPTO 2003*, pages 126–144. Springer-Verlag, 2003. Leture Notes in Computer Science No. 2729.
9. J. Kilian and E. Petrank. Identity escrow. In *Proc. CRYPTO 98*, pages 169–185. Springer-Verlag, 1998. Lecture Notes in Computer Science No. 1642.
10. J. Liu, V. Wei, and D. Wong. Custodian-hiding verifiable encryption. In *WISA 2004*, pages 54–67. Springer-Verlag, 2004. Lecture Notes in Computer Science No. 3325.
11. J. Liu, V. Wei, and D. Wong. Linkable spontaneous anonymous group signature for ad hoc groups. In *ACISP04*, pages 325–335. Springer-Verlag, 2004. Lecture Notes in Computer Science No. 3108.

12. K. Ohta and T. Okamoto. On concrete security treatment of signatures derived from identification. In *Proc. CRYPTO 98*, pages 354–369. Springer-Verlag, 1998. LNCS Vol. 1462.
13. P. Paillier. Public-key cryptosystems based on composite residuosity classes. In *Proc. EUROCRYPT 99*, pages 223–239. Springer-Verlag, 1999. Lecture Notes in Computer Science No. 1592.
14. D. Pointcheval and J. Stern. Security proofs for signature schemes. In *Proc. EUROCRYPT 96*, pages 387–398. Springer-Verlag, 1996. LNCS Vol. 1070.
15. R. Rivest, A. Shamir, and Y. Tauman. How to leak a secret. In *Proc. ASIACRYPT 2001*, pages 552–565. Springer-Verlag, 2001. Lecture Notes in Computer Science No. 2248.
16. M. Stadler. Publicly verifiable secret sharing. In *Proc. EUROCRYPT 96*, pages 191–199. Springer-Verlag, 1996. Lecture Notes in Computer Science No. 1070.

A   Proof of Theorem 1

Proof. Correctness of our proposed scheme is trivial and its proof is thus omitted. The proof of the theorem is then a direct implication of the three lemmas that follow.   □

Lemma 1 (Soundness). *Our proposed scheme is sound in the random oracle model if the Strong RSA assumption holds.*

Proof. Assume there is a PPT algorithm $\mathcal{P}^*$, which can produce a ciphertext ψ (corresponding to δ) with non-negligible probability such that $\mathcal{V}$ outputs accept but no one can decrypt, or compute m' such that $(m', \delta) \in \mathcal{R}$. That is,

$$\Pr[(\mathcal{D}(\mathsf{param}, N, \{\mathsf{PK}\}_N, \pi, \mathsf{SK}_\pi, \psi, L), \delta) \notin \mathcal{R}] > \mathsf{neg}(\lambda)$$

for at least one $\pi \in \{1, \ldots, N\}$. for some PPT algorithm $\mathcal{V}$ and $\mathcal{D}$.

We construct a PPT simulator (the reduction master) $\mathcal{M}$ which has N private keys $\mathsf{SK}_1, \ldots, \mathsf{SK}_N$ and calls $\mathcal{P}^*$ to compute an integer x such that $(x, \delta) \in \mathcal{R}$.

$\mathcal{M}$ also controls the random oracle H. It flips coins for H and records queries to the oracle. It maintains the consistnecy of H. $\mathcal{P}^*$ is allowed the query the random oracle at most q_H times.

$\mathcal{P}^*$ generates a ciphertext ψ (corresponding to δ), consists of c_1, $(u_1, e_1, v_1, t_1, \tilde{r}_1, \tilde{s}_1, \tilde{m}_1)$, $\ldots$, $(u_N, e_N, v_N, t_N, \tilde{r}_N, \tilde{s}_N, \tilde{m}_N)$ where it satisfies the verification including the following N equations: $c_{i+1} = H(L, \delta_i, \hat{u}_i, \hat{e}_i, \hat{v}_i, \hat{\delta}_i, \hat{t}_i)$ for $i = 1, \ldots,$ $N-1$ and $c_1 = H(L, \delta_N, \hat{u}_N, \hat{e}_N, \hat{v}_N, \hat{\delta}_N, \hat{t}_N)$ where $\hat{u}_i = u_i^{2c_i} g_i^{2\tilde{r}_i}$, $\hat{e}_i = e_i^{2c_i} y_{1_i}^{2\tilde{r}_i} h_i^{2\tilde{m}_i}$, $\hat{v}_i = v_i^{2c_i} (y_{2_i} y_{3_i}^{H(u_i, e_i)})^{2\tilde{r}_i}$ $\hat{\delta}_i = \delta_i^{c_i} \gamma_i^{\tilde{m}_i}$, $\hat{t}_i = t_i^{c_i} \mathsf{g}_i^{\tilde{m}_i} \mathfrak{h}_i^{\tilde{s}_i}$ for $i = 1, \ldots, N$

The master $\mathcal{M}$ will invoke $\mathcal{A}$ with constructed inputs, receive and process outputs from $\mathcal{A}$, and may invoke $\mathcal{P}^*$ for multiple times depending on $\mathcal{P}^*$'s outputs from previous invocations. In the random oracle model, $\mathcal{M}$ flips the coins for the random oracle H record queries to the oracle. Consider each invocation of $\mathcal{P}^*$ to be recorded on a simulation transcript tape. Some transcripts produce successful ciphertext. Others do not.

Let $\mathbf{E}$ be the event that each of the N queries corresponding to the N Verification queries have been included in the q_H queries $\mathcal{P}^*$ made to the random oracles. In the event $\bar{\mathbf{E}}$, $\mathcal{M}$ needs to flip additional coins in order to verify $\mathcal{P}^*$'s ciphertext. Then the probability of c_1 satisfying the (final) Verification equation is at most $1/(2^k - q_H)$ because $\mathcal{P}^*$ can only guess the outcomes of queries used in Verification that he has not made. Therefore

$$\mathsf{neg}(\lambda) < \Pr[\mathbf{E}]\Pr[\mathcal{P}^* \text{ succeed}|\mathbf{E}] + \Pr[\bar{\mathbf{E}}]\Pr[\mathcal{P}^* \text{ succeed}|\bar{\mathbf{E}}]$$

$$\leq \Pr[\mathbf{E}]\Pr[\mathcal{P}^* \text{ succeed}|\mathbf{E}] + 1 \cdot \left(\frac{1}{2^k - q_H}\right)$$

and

$$\Pr[\mathbf{E} \text{ and } \mathcal{P}^* \text{ succeed}] > \mathsf{neg}(\lambda) - \left(\frac{1}{2^k - q_H}\right)$$

Hence the probability of $\mathcal{P}^*$ returning a valid ciphertext and having already queried the random oracle for all the N queries used in Verification is essentially greater than $\mathsf{neg}(\lambda)$ as $\frac{1}{2^k - q_H}$ is negligibly small.

Therefore, in each $\mathcal{P}^*$ transcript which produced a valid ciphertext, there exists N queries to H, denoted by $X_{i_1}, \cdots, X_{i_N}, 1 \leq i_1 < \cdots < i_N$, such that they match the N queries made in Verification. This happens with each transcript that $\mathcal{P}^*$ successfully produces a valid ciphertext, with negligible exceptions.

In creating a successful ciphertext ψ by $\mathcal{P}^*$, consider the set of all queries made by $\mathcal{P}^*$ that were used (including duplicate queries) in Verification. Let $X_{i_1}, \cdots, X_{i_N}$ denote the first appearance of each of the queries used in Verification, $i_1 < \cdots < i_N$. Let π be such that $X_{i_N} := H(L, \delta_{\pi-1}, \hat{u}_{\pi-1}, \hat{e}_{\pi-1}, \hat{v}_{\pi-1}, \hat{\delta}_{\pi-1}, \hat{t}_{\pi-1})$ in Verification. We call π the *gap* of ψ.

We call a successful creation of ψ by $\mathcal{P}^*$ a (ℓ, π)-ψ if $i_1 = \ell$. That is, the first appearance of all Verification-related queries is the ℓ-th query and the gap equals π. There exist ℓ and π, $1 \leq \ell \leq q_H$, $1 \leq \pi \leq N$, such that the probability $\mathcal{P}^*$ produces (ℓ, π)-ψ is no less than $1/(Nq_H\mathsf{neg}(\lambda))$.

In the following, $\mathcal{M}$ will do a rewind-simulation for each value of ℓ and π.

In the rewind-simulation for a given (ℓ, π), $\mathcal{M}$ first invokes $\mathcal{P}^*$ to obtain its output and its Turing transcript $\mathcal{T}$. $\mathcal{M}$ computes the output and the transcript to determine whether they form a successful (ℓ, π)-ψ. If not, abort. Otherwise continue. This can be done in at most polynomial time because $\mathcal{M}$ records queries made by $\mathcal{P}^*$ to the random oracles. The transcript $\mathcal{T}$ is rewound to the ℓ-th query and given to $\mathcal{P}^*$ for a rewind-simulation to generate transcript $\mathcal{T}'$. New coin flips independent of those in $\mathcal{T}$ are made for all queries subsequent to the ℓ-th query while maintaining consistencies with the prior queries. $\mathcal{T}$ and $\mathcal{T}'$ use the same code in $\mathcal{P}^*$. The ℓ-th query, common to $\mathcal{T}$ and $\mathcal{T}'$, is denoted $c_{\pi+1} = H(L, \delta_\pi, u'_\pi, e'_\pi, v'_\pi, \gamma^u_\pi, t'_\pi)$ $\mathcal{M}$ knows γ^u_π but not u at the time of the rewind. After $\mathcal{P}^*$ returns the output from the rewind simulation, $\mathcal{M}$ proceeds to compute the DL of δ_π, that is, u.

By the forking lemma [14], heavy-row lemma [12] or the Rewind-on-Success lemma [11], there exists non-negligible probability that $\mathcal{P}^*$ produces two (ℓ, π)-ψ

from the tape $\mathcal{T}$ and a rewind-simulation tape $\mathcal{T}'$ with $\gamma_\pi^u = \gamma_\pi^{\tilde{m}_\pi + c_\pi m}$ from $\mathcal{T}$ and $\gamma_\pi^u = \gamma_\pi^{\tilde{m}'_\pi + c'_\pi m}$ from $\mathcal{T}'$

Solve for the equations to obtain m. Using the argument in the proof of Theorem 4 in [7] (and by the Strong RSA Assumption), $(m, \delta) \in \mathcal{R}$. That is, ψ is an encryption of the witness of δ. Desired contradiction occurs.

Lemma 2 (Zero-knowledge). *Our proposed scheme is zero-knowledge in the random oracle model.*

Proof. (Sketch.) This is rather obvious due to the symmetry enjoyed by the ring-structure of the ciphertext validity proof. □

Lemma 3 (Anonymity). *Our proposed scheme is anonymous in the random oracle model if DCR assumption holds.*

Proof. (Sketch.) Observe that $(u_i, e_i, v_i, t_i, \tilde{r}_i, \tilde{s}_i, \tilde{m}_i), i = 1, \dots, n, i \neq \pi$ are all random numbers chosen uniformly. At the closing point, $(u_\pi, e_\pi, v_\pi, t_\pi, \tilde{r}_\pi, \tilde{s}_\pi, \tilde{m}_\pi)$ also distribute uniformly since r_π and s_π are uniformly chosen from $[n_\pi/4]$. Remaining c_1 is the output of a hash function which can be regarded as a random number as well. □

SKiMPy: A Simple Key Management Protocol for MANETs in Emergency and Rescue Operations[*]

Matija Pužar[1], Jon Andersson[2], Thomas Plagemann[1], and Yves Roudier[3]

[1] Department of Informatics, University of Oslo, Norway
{matija, plageman}@ifi.uio.no
[2] Thales Communications, Norway
jon.andersson@no.thalesgroup.com
[3] Institut Eurécom, France
yves.roudier@eurecom.fr

Abstract. Mobile ad-hoc networks (MANETs) can provide the technical platform for efficient information sharing in emergency and rescue operations. It is important in such operations to prevent eavesdropping, because some the data present on the scene is highly confidential, and to prevent induction of false information. The latter is one of the main threats to a network and could easily lead to network disruption and wrong management decisions. This paper presents a simple and efficient key management protocol, called SKiMPy. SKiMPy allows devices carried by the rescue personnel to agree on a symmetric shared key, used primarily to establish a protected network infrastructure. The key can be used to ensure confidentiality of the data as well. The protocol is designed and optimized for the high dynamicity and density of nodes present in such a scenario. The use of preinstalled certificates mirrors the organized structure of entities involved, and provides an efficient basis for authentication. We have implemented SKiMPy as a plugin for the Optimized Link State Routing Protocol (OLSR). Our evaluation results show that SKiMPy scales linearly with the number of nodes in worst case scenarios.

1 Introduction

Efficient collaboration between rescue personnel from different organizations is a mission critical element for a successful operation in emergency and rescue situations. There are two central requirements for efficient collaboration, the incentive to collaborate, which is naturally given for rescue personnel, and the ability to efficiently communicate and share information. Mobile ad-hoc networks (MANETs) can provide the technical platform for efficient information sharing in such scenarios, if the rescue personnel is carrying and using mobile computing devices with wireless network interfaces.

Wireless communication needs to be protected to prevent eavesdropping. The data involved should not be available to any third parties, for neither publication or mali-

---

[*] This work has been funded by the Norwegian Research Council in the IKT-2010 Program, Project Nr. 152929/431. It has been also partly supported by the European Union under the E-Next SATIN-EDRF project.

R. Molva, G. Tsudik, and D. Westhoff (Eds.): ESAS 2005, LNCS 3813, pp. 14–26, 2005.
© Springer-Verlag Berlin Heidelberg 2005

cious actions. Another important requirement is to prevent inducing of false data. At the application layer this might for example lead to wrong management decisions. At the network layer it has been shown that a very few percent of misbehaving nodes easily can lead to network disruption and partitioning [17]. In both cases, efficiency of the rescue operation will be drastically reduced and might ultimately cause loss of human lives. In order to prevent such a disaster, all data traffic should be protected, allowing only authorized nodes access to the data. Given that devices carried by the rescue personnel will mostly have limited resources, any security scheme based solemnly on asymmetric cryptography will be too costly in terms of computing power, speed and battery consumption. Therefore, the use of symmetric encryption with shared keys is preferable for MANETs in emergency and rescue scenarios. Agreeing on a shared key in a highly dynamic and infrastructure-less MANET is a non-trivial problem and requires establishing trust relations between all devices. It is important for emergency and rescue scenarios that corresponding solutions are simple, efficient, robust, and autonomous. User interactions should be kept at an absolute minimum.

This paper describes a simple key management protocol, called SKiMPy, that can be used to establish a symmetric shared key between the rescue personnel's devices. By this, SKiMPy will set up a secure network infrastructure between authorized nodes, while keeping out unauthorized ones. It may be decided at the application layer whether the established shared key is robust enough for achieving some degree of data confidentiality as well. The basis for this simple and efficient solution is the fact that rescue personnel are members of public organizations with strict, well defined hierarchies. This hierarchy can be mirrored into a certificate structure installed a priori on their devices, i.e., before the accident or disaster actually happens. As a result, it is possible for the nodes during the rescue activity to authenticate each other on a peer-to-peer basis, without need for contacting a centralized server or establishing trust in a distributed approach.

The organization of the paper is as follows. Section 2 gives a detailed description of our protocol. In Section 3 we show some design considerations and respective solutions. Section 4 describes an implementation of the protocol together with evaluation results. In Section 5 we present related work. Finally, conclusion and future work are given in Section 6.

2 Protocol Description

SKiMPy makes use of the existing traffic in the network to trigger key exchange. Periodic routing beacons (HELLO), sent by proactive routing protocols, are such an example. The following two messages are specific to SKiMPy:

- *Authentication Request* (AUTH_REQ): sent by a node after it detects traffic from a node having a key that is *worse* than its own one. The message is used to inform the remote node that the sending node is willing to transfer its key.
- *Authentication Response* (AUTH_RESP): sent by a node, as a result of a received AUTH_REQ message. The message is used to inform the remote party that the node is willing to perform the authentication and receive the remote and *better* key.

The protocol consists of three phases, namely *(I) Neighborhood Discovery, (II) Batching* and *(III) Key Exchange.*

During phase I, a node listens to all traffic sent by its immediate neighbors. If it detects a node using a *worse* key (explained in detail in Section 3.2), it will send an *Authentication Request* message to it, saying it is willing to pass on its key. Upon receiving such a message, the other node enters the phase II, waiting for possible other authentication requests before sending a response. This batching period is used for optimization - a node will only perform authentication with the *best* of all neighbors. All the other keys will, due to the transitiveness property of the *better than* relation, at some point get overruled and therefore there is no point in getting them. After the node has chosen its peer, it sends an *Authentication Response* after which its peer initializes the actual authentication procedure, that is, exchange of certificates, establishing a secure tunnel, and finally transfer of the key. The reason for having such a handshake procedure is to ensure that the nodes can indeed communicate. In some standards, such as 802.11b [19], traffic like broadcast messages can be sent on a lower transmitting rate with larger transmission range than data messages. Thus, broadcast messages might reach a remote node and trigger a key exchange, even though the nodes cannot directly exchange data packets.

Figure 1 shows an example of the key exchange between three nodes (A, B and C) and indicates the different phases of the key exchange for node A. Node A enters phase I when turned on. Nodes B and C do not directly hear each other's traffic and are only able to communicate through node A, once the shared key is fully deployed.

The initial states of the three nodes are as follows: A has the key K_A, B has K_B and C has K_C. In this example, K_C is the *best* key, whereas K_A is the *worst* key.

Phase I:
1. Node A is turned on. All nodes send periodic HELLO messages which are part of the routing protocol.
2. A receives a HELLO message from B, notices a key mismatch, but ignores it because K_A is *worse* than K_B.
3. A receives HELLO from C, notices a key mismatch, but ignores it because K_A is *worse* than K_C.
4. B and C receive HELLO from A, they both notice they have a *better* key than K_A, and after a random time delay (to prevent traffic collisions), send an AUTH_REQ message to A.

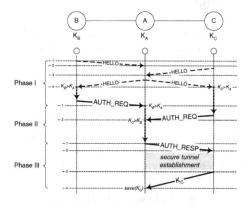

Fig. 1. Message Flow Diagram

Phase II:
1. *A* receives AUTH_REQ from *B* notices that *B* has a *better* key and schedules authentication with *B*. The authentication is to be performed after a certain waiting period, in order to hear if some of the neighbors has an even *better* key.
2. *A* receives AUTH_REQ from *C* as well, sees that *C* has a key *better* than K_B, and therefore decides to perform authentication with *C* instead.

Phase III:
1. *A* sends an AUTH_RESP message to *C*, telling it is ready for the authentication process
2. *C* initiates the authentication procedure with *A*, they exchange and verify certificates; the secure tunnel is established.
3. *C* sends its key K_C to *A* through the secure tunnel.
4. *A* receives the key and saves it locally; the old key K_A is saved in the key repository for eventual later use; *A* sends the new key further, encrypted with K_A.

In the next round, that is, after it hears traffic from node *B* signed with K_B, node *A* will use the same procedure to deliver the new key K_C to node *B*, hence establishing a common shared key in the whole cell.

There are two important parameters which influence the performance of the protocol and therefore have to be chosen carefully. The delays used before sending AUTH_REQ are random, to minimize the possibility of collisions in the case when more nodes react to the same message. On the other hand, the delay from the moment a node receives AUTH_REQ to the moment it chooses to answer with AUTH_RESP is a fixed interval and should be tuned so that it manages to hear as many neighbors as possible within a reasonable time limit. By this, all nodes that have been heard during the waiting period can be efficiently handled in the same batch.

3 Design Considerations

Our protocol is designed for highly dynamic networks, where nodes may appear, disappear and move in an arbitrary manner. Topology changes are inevitable. The key management protocol must have low impact on the available resources, i.e. battery, bandwidth and CPU time. Here, we analyze the different security and performance issues that had to be considered while designing the protocol, as well as respective solutions integrated into SKiMPy.

3.1 Authentication

An important characteristic of an emergency and rescue operation is that the organizations involved (police, fire department, paramedics, etc.) are often well structured, public entities. Before the rescue personnel comes to the disaster scene, all devices are prepared for their tasks. One task in the preparation phase, which we call *a priori* phase [23], is the installation of valid certificates. The certificates are signed by a commonly trusted authority, such as the ministry of internal affairs, ministry of defense, etc., on the top of the trust chain. This gives nodes the possibility to authenticate each other without need for contacting a third party.

Certificates on the nodes can identify devices, users handling them, or even both. The users would then present their certificate to the device by means of a token, i.e. smartcard. The decision for this does not impact the key management in SKiMPy, but it impacts the way how lost and stolen nodes are handled, i.e., revoking certificates and/or blacklisting of such nodes. We explain this issue later, in Section 3.5.

3.2 Choosing Keys

The main task of SKiMPy is to make sure that all the nodes agree on a shared key. When a node is turned on, it generates a random key with a random ID number. The uniqueness of the key IDs must be ensured by e.g. using the hash value of the key itself as part of the ID, by including the nodes MAC address, etc. The final shared key is always chosen from nodes' initial keys. To achieve this, we introduce the notions of *better* and *worse* keys, together with the relation ">" representing *better than*. There are several possible schemes for deciding which of the keys is *better* or *worse* and all schemes can be equally valid, as long as they cannot cause key exchange loops, are unambiguous and transitive: $(A > B$ and $B > C) => A > C$. The necessary control information, which depends on the scheme chosen, is always sent with the message signature.

We briefly describe two schemes and their advantages and drawbacks.

The first scheme uses arithmetic comparison of two numbers, i.e. the key having a higher or lower ID number, timestamp or a similar parameter, is considered to be *better*. The advantage of this scheme is that it is unambiguous, transitive and easy to implement. In addition, it can be "tweaked" in a way that would prevent a single node to cause re-keying of an already established network cell. For example, if the scheme defines that the lower ID number means a better key, the highest bit of the ID number can be always set to "1" when the node is turned on, and cleared once two nodes merge. Assuming that nodes in a certain area will in most cases pop up independently, this simple and yet efficient method might prevent a lot of unnecessary re-keying traffic. If we use the keys' timestamps instead of the ID numbers, choosing a lower timestamp could imply that the key is older and that more nodes have it already. SKiMPy does not require the clocks of different devices to be synchronized and therefore, the given assumption might not necessarily be true, especially if the key creator's clock was heavily out of sync. One major drawback of the presented scheme is that a small cell (consisting of, for example, 2 nodes) could easily cause re-keying of a much bigger cell (having, for example, 100 nodes), which would be a waste of resources.

The second scheme takes care of this problem by using the number of nodes in each network cell as the decisive factor. The simple rule for this scheme is to always re-key the smaller cell, i.e. the one with the lower number of nodes, thus minimizing resource consumption for the necessary re-keying. The approximate number of nodes can be either retrieved from the routing protocol state information (if, for example, the OLSR routing protocol [7] is used) or maintained at a higher protocol layer, as it is done in our project. However, if not all of the nodes have exactly the same information (which is to be expected in a dynamic scenario), and for some obscure reason we have more simultaneous merging processes between the same two cells, a key exchange loop may occur. One approach to this problem is to adjust in each node the state information of the number of nodes in its cell, always increasing it when new nodes join, but never decreasing it upon partitioning of the cell.

At the present, we use the first scheme, choosing always a key with a lower ID number. An in-depth study of both schemes and their variations is subject to ongoing and future work.

3.3 Key Distribution

Once a node gets a new key as a result of network merging, the key should be deployed within its previous network cell. There are several ways to achieve this:

- *Proactively* - each node receiving the key immediately forwards it to the others. This approach ensures prompt delivery of the key to all nodes, but it also generates a lot of unnecessary network traffic.
- *Reactively* - when a node receives a key, it does nothing. Only after detecting a message sent by a neighbor and signed with the old key, the node sends the new key further. This approach uses less resources, but it takes more time for the whole cell to get a stable key.
- *Combination* - the first node getting the new key (that is, the node which performed the merge) immediately forwards the key to its one-hop neighbors, since it knows that no other node in its previous cell has it yet. The other nodes do not distribute it right away, but rather when (if) they notice that a node still uses an old key. This approach keeps the number of necessary broadcast messages containing the key at a minimum.

In any of the given cases, the new key is encrypted using the old one before sending, giving all the other nodes the possibility to immediately start using it. The old key is saved for a short period of time, for possible latecomers. This can be done because in this particular case the key change was not performed explicitly for the purpose of preventing traffic analysis attacks.

In our implementation, described in Section 4.1, we use the *combination* approach.

3.4 Key Update

When created, each key has a companion key (called *update key*) used to periodically update it. The update key is never used on traffic that goes onto the network and therefore it is not prone to traffic-analysis attacks. The nodes must periodically update the main key. The new key can be computed using one-way hash functions such as SHA-1 [15] or MD5 [25], ensuring backward secrecy in the case the key gets broken at some stage. In addition to the ID of the key used to sign it, a message contains also the update-number saying how many times the key on the sender-node has been updated. That way, the receiver can easily compute the new key if it notices a mismatch, which could happen since we can't expect all the nodes to perform the update at exactly the same time. The local update will not take place if the received message has an invalid signature.

3.5 Exclusion of Nodes

Once authenticated, a node is a fully trusted member of the network. This poses the evident problem of how to exclude such a node once the device has been lost or, even worse, stolen by a malicious third party. At the present, exclusion of already authenticated nodes is not solved in SKiMPy and is part of ongoing and future work. Here, we describe some ideas on measures to be taken in order to ensure that such a node stays out of the network.

First, the node's certificate must be revoked, preventing the node from re-authenticating later at some stage. Since there is no central authority, a decision is reached on which node or person can perform the task of revoking certificates. If the certificates contain also additional attributes such as rank or role of the persons (assuming that the certificates do in fact represent persons, not devices), it can be decided that only certain roles/ranks (such as *leader*) can perform revocation and blacklisting. In theory, the leaders' devices might also be stolen, but in practice they should normally be physically well protected. It is important to ensure that the compromised node itself does not revoke and blacklist legitimate ones or, even worse, the whole network.

Next, the node's IP address should be put on a common blacklist. Assuming that IP addresses are bound to the certificates (as presented in e.g. [22]), the nodes would be unable to change their IP address. However, relying on fixed IP addresses might introduce new issues and should be considered carefully. Traffic coming from blacklisted nodes must be discarded at the lowest possible layer and, in case legally signed traffic coming from a blacklisted node is detected, the compromised key must be removed.

Additional methods might be used to ensure that devices cannot be used by unauthorized persons. One such example is a system relying on short range wireless authentication tokens. A token is installed into the personnel's vests or watches, ensuring confidentiality of the data and denying unauthorized access to the devices when they get out of their token's range [8].

3.6 Batching

To save resources as much as possible, our protocol makes the nodes learn about their neighborhood before acting, reducing the number of performed authentications and thus reducing directly CPU and bandwidth consumption. This is possible due to the fact that all nodes directly trust the same certificate authority and, therefore, if a node has been successfully authenticated before and has received the shared secret, we implicitly trust it.

Emphasis has been put on optimization with regards to number of messages sent out in the air. We measured the number of certificates and key management messages exchanged, and compared these figures to the number of routing messages needed from the moment when the nodes were turned on, up to the moment when a stable shared key was established. To perform these measurements, we used a static, wired test bed with 16 nodes.

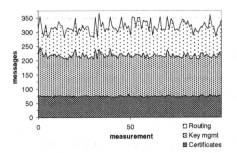
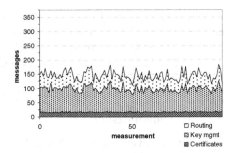

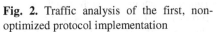

Fig. 2. Traffic analysis of the first, non-optimized protocol implementation

Fig. 3. Results for the same scenario, after introducing the batching process

Figures 2 and 3 show that introducing neighborhood awareness approximately halved the total number of messages and, proportionally, the time needed to reach a stable state. Moreover, the number of messages carrying certificates, whose size is much larger than other key management messages, has been reduced to approximately 23% of the initial number. The authentication was considered to be done after the exchange of certificates. Therefore, the results shown here are only an approximation, and might be slightly different when an actual authentication algorithm is used.

3.7 Additional Issues

The protocol's goal is to establish a secure network infrastructure. SKiMPy makes it impossible for a misbehaving node to induce a key that has either expired, or that would not have been selected in a normal operation. Such keys will be immediately discarded.

Timeouts are used during the *Key Exchange* phase (explained in Section 2) to ensure that a node does not end up in indefinite wait states or deadlocks as a result of possible link failures. Care must be taken for possible Denial-of-Service attacks in any of these cases.

In the *closing* phase of the rescue operation [23], the keys must be removed to prevent them from being possibly reused afterwards on a different rescue site.

4   Protocol Implementation and Evaluation

4.1   Implementation

Optimized Link State Routing Protocol (OLSR) [7] is a proactive routing protocol for ad-hoc networks which is one of the candidates to be used in our solution for the emergency and rescue operations. The olsr.org OLSR daemon [28] is the implementation we decided to test, since it is portable and expandable by means of loadable plugins. One example of such a plugin, present in the main distribution, is the Secure OLSR plugin [16]. The plugin is used to add signature messages to OLSR traffic, only allowing nodes that possess the correct shared (pre-installed) key to be part of the OLSR routing domain. One important functionality this plugin lacks is a key management protocol. Even though SKiMPy is mainly designed to protect all traffic and not only routing, it is still a good opportunity to test and analyze it in a realistic environment with a real routing protocol.

The key management protocol has been coded directly into the security plugin, although the plans are to make it as a separate one. X.509 certificates [18] and OpenSSL [27] are currently used to perform node authentication.

4.2 Evaluation Results

To facilitate development of this and other protocols, we created an emulation test bed, called NEMAN [24]. Routing daemons run independently, each attached to a different virtual Ethernet device. We use the monitoring channel of the emulator to analyze the keys used by each of the routing daemons. In order to test performance and scalability the protocol, we have made measurements from 2 to 100 nodes, with two very different kinds of scenario: chain and mesh. Figures 4 and 5 show example screenshots taken from the GUI, representing the two different scenarios.

In a chain scenario, the nodes are lined up in a single chain and the distance between all nodes in the chain is such that only the direct neighbors can communicate in a single hop with each other. We consider this to be the worst case scenario still giving full network connectivity. Given that all the nodes have to perform authentication with both their neighbors, this leaves no place for optimization, i.e. batching during the waiting period.

In a mesh scenario, however, nodes have multiple, randomly scattered neighbors, as it is natural in ad-hoc networks. Having multiple neighbors allows the protocol to exploit the batching phase, reducing traffic and resource consumption.

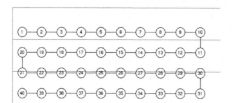

Fig. 4. Example of a chain scenario

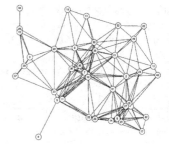

Fig. 5. Example of a mesh scenario

Ten independent runs were performed for each number of nodes and each scenario. All the nodes were started simultaneously (which we assume is the worst case for our protocol), with a random key and key ID. To be able to meaningfully compare the results, the nodes were static and the density was constant. The delay in the batching period was set to be 1 second, i.e. half of the interval used by OLSR to send HELLO messages.

One important fact that the results on Figure 6 immediately show is that the protocol scales linearly with linear increase of the number of nodes and physical network area accordingly (thus giving the same density of nodes). After approximately 10 nodes, the total time became almost independent on the network size. By the fourth second, most authentications have already been performed and the key distribution process came into place. In some additional measurements, we introduced node movement using the random waypoint mobility model. As long as all of the nodes remained reachable and the density was constant, movement did not induce a notable delay.

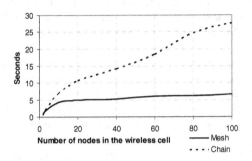

Fig. 6. Time needed to achieve a stable shared key

We also proved that having multiple neighbors does in fact lower the time necessary to reach a stable state. This scenario gives less deviation as well, which is understandable since in the case of chain there is more fluctuation of keys, nicely seen in the GUI.

5 Related Work

Different authentication schemes are available as a starting point for key management.

Devices can exchange a secret or pre-authentication data through a physical contact or directed infrared link between them [3, 26]. Another way is for the users to compare strings displayed on their devices (a representation of their public key, distance between them, etc. as presented in [9]). Since user interaction in a rescue operation should be kept as minimum, we need a different approach.

Threshold cryptography schemes, such as [20] and [31] require all nodes that are going to perform signatures to carry a share of the group private key. The full signature is acquired by a certain, predefined number of nodes who present partial signatures computed using their shares. These schemes allow a small number of nodes to be compromised and still not to present a threat for the network. However, since we do not know the number of nodes that can be expected at the rescue scene and small partitions might always be present, this approach is not suited for our scenario.

Čapkun et al. [10] present a fully self-organized public-key management system that does not rely on trusted authorities, developed mainly for networks where users can join and leave without any centralized control. This is not applicable to networks used in rescue operations, where only authorized nodes are allowed to participate. In [11], they present a solution similar to ours, explained in Section 3.1, allowing nodes to authenticate each other by means of pre-installed certificates with a common authority. The advantages of such a system are twofold: first, the data in the network is more secure. Second, establishing trust and agreeing on a shared key is much more efficient, i.e., faster and less resources are consumed.

Related key management protocols can be roughly divided into the following three categories [6].

The first one relies on a fixed infrastructure and servers that are always reachable. Since we never know where accidents will happen, we should expect them to happen at places where we cannot rely on the fact that fixed infrastructure will be present.

The next category comprises contributory key agreement protocols, which are not suited for our scenario for several reasons. Such protocols ([1, 5, 12, 29, 30], to name a few) are based on Diffie-Hellman two-party key exchange [13] where all the nodes give their contribution to the final shared key, causing re-keying every time a new node joins or an existing node leaves the group. In an emergency and rescue operation, we can expect nodes to pop up and disappear all the time, often causing network partitioning and merging. Therefore, using contributory protocols would cause a lot of computational and bandwidth costs which cannot be afforded. Besides, most of these protocols rely on some kind of hierarchy (chain, binary tree, etc.) and a group manager to deploy and maintain shared keys. In a highly dynamic scenario this approach would be quite ineffective. Another reason why such protocols are not suited for us, is that in order for the nodes to be able to exchange keys, a fully working routing infrastructure has to be established prior to that. Since the routing protocol is one of the main things we need to protect, this is a major drawback. Asokan and Ginzboorg [2] present a password-based authenticated key exchange system. A weak password is known to every member and it is used by each of them to compute a part of the final shared key. This approach shares some already mentioned drawbacks and introduces new ones which conflict with our scenario and requirements. User interaction is needed and it is assumed that all the members are present when creating the key.

The last category are protocols based on key pre-distribution. The main characteristic of such protocols is that a pair or group of nodes can compute a shared key out of pre-distributed sets of keys present on each node. These sets of keys are either given by a trusted entity before the nodes come to the scene [4, 14, 21], or chosen and managed by the nodes themselves, as it is done in DKPS [6].

SKiMPy is different in the sense that it uses pre-installed certificates to perform direct authentication between two nodes. This makes it more simple and efficient.

6   Conclusion

In this paper, we presented a simple and efficient key management protocol, called SKiMPy, developed and optimized especially for highly dynamic ad-hoc networks. The protocol relies on the fact that there will be an *a priori* phase of rescue and emergency operations, within which certificates will be deployed on rescue personnel's devices. Pre-installed certificates are necessary due to the fact that highly sensitive data may be exchanged between the rescue personnel. The certificates make it possible for the nodes to authenticate each other without need for a third party present on the scene.

We described a proof-of-concept implementation, as well as evaluation results. The results show that SKiMPy performs very well and it scales linearly with the number of nodes. As part of further work we will analyze more in-depth different key selection and distribution schemes, authentication protocols, and fine tune certain protocol parameters, like the delays described in Section 2. Open issues like exclusion of compromised nodes, duplicate key ID numbers, denial of service attacks, etc. are also subject of further investigation.

References

1. Alves-Foss, J., "An Efficient Secure Authenticated Group Key Exchange Algorithm for Large And Dynamic Groups", Proceedings of the 23rd National Information Systems Security Conference, pages 254-266, October 2000
2. Asokan, N., Ginzboorg, P., "Key Agreement in Ad Hoc Networks", Computer Communications, 23:1627-1637, 2000
3. Balfanz, D, Smetters, D. K., Stewart, P, Wong, H. C., "Talking To Strangers: Authentication in Ad-Hoc Wireless Networks", Proceedings of the 9th Annual Network and Distributed System Security Symposium (NDSS'02), San Diego, California, February 2002
4. Blom, R., "An Optimal Class of Symmetric Key Generation System", Advances in Cryptology - Eurocrypt'84, LNCS vol. 209, p. 335-338, 1985
5. Bresson, E., Chevassut, O., Pointcheval, D., "Provably Authenticated Group Diffie-Hellman Key Exchange - The Dynamic Case (Extended Abstract)", Advances in Cryptology - Proceedings of AsiaCrypt 2001, pages 290-309. LNCS, Vol. 2248, 2001
6. Chan, Aldar C-F., "Distributed Symmetric Key Management for Mobile Ad hoc Networks", IEEE Infocom 2004, Hong Kong, March 2004
7. Clausen T., Jacquet P., "Optimized Link State Routing Protocol (OLSR)", RFC 3626, October 2003
8. Corner, Mark D., Noble, Brian D., "Zero-Interaction Authentication", at The 8th Annual International Conference on Mobile Computing and Networking (MobiCom'02), Atlanta, Georgia, September 2002
9. Čagalj, M., Čapkun, S., Hubaux, J.-P., "Key agreement in peer-to-peer wireless networks", to appear in Proceedings of the IEEE (Specials Issue on Security and Cryptography), 2005
10. Čapkun, S., Buttyán, L., Hubaux, J.-P., "Self-Organized Public-Key Management for Mobile Ad Hoc Networks", IEEE Transactions on Mobile Computing, Vol. 2, No. 1, January-March 2003
11. Čapkun, S., Hubaux, J.-P., Buttyán, L., "Mobility Helps Security in Ad Hoc Networks", In Proceedings of the 4th ACM Symposium on Mobile Ad Hoc Networking and Computing (MobiHoc'03), Annapolis, Maryland, June 2003
12. Di Pietro, R., Mancini, L., Jajodia, S., "Efficient and Secure Keys Management for Wireless Mobile Communications", Proceedings of the second ACM international workshop on Principles of mobile computing, pages 66-73, ACM Press, 2002
13. Diffie, W., Hellman, M., "New directions in cryptography", IEEE Transactions on Information Theory, 22(6):644-652, November 1976
14. Eschenauer L., Gligor, Virgil D., "A Key-Management Scheme for Distributed Sensor Networks", Proceedings of the 9th ACM Conference on Computer and Communication Security (CCS'02), Washington D.C., November 2002
15. Federal Information Processing Standard, Publication 180-1. Secure Hash Standard (SHA-1), April 1995
16. Hafslund A., Tønnesen A., Rotvik J. B., Andersson J., Kure Ø., "Secure Extension to the OLSR protocol", OLSR Interop Workshop, San Diego, August 2004

17. Hollick, M., Schmitt, J., Seipl, C., Steinmetz, R., "On the Effect of Node Misbehavior in Ad Hoc Networks", Proceedings of IEEE International Conference on Communications, ICC'04, Paris, France, volume 6, pages 3759-3763. IEEE, June 2004

18. Housley, R., Ford, W., Polk, W. and D. Solo, "Internet X.509 Public Key Infrastructure", RFC 2459, January 1999

19. IEEE, "IEEE Std. 802.11b-1999 (R2003)", http://standards.ieee.org/getieee802/download/802.11b-1999.pdf

20. Luo, H., Kong, J., Zerfos, P., Lu, S., Zhang, L., "URSA: Ubiquitous and Robust Access Control for Mobile Ad-Hoc Networks", IEEE/ACM Transactions on Networking, October 2004

21. Matsumoto, T., Imai, H., "On the key predistribution systems: A practical solution to the key distribution problem", Advances in Cryptology - Crypto'87, LNCS vol. 293, p. 185-193, 1988

22. Montenegro, G., Castelluccia, C., "Statistically Unique and Cryptographically Verifiable (SUCV) Identifiers and Addresses", NDSS'02, February 2002

23. Plagemann, T. et al., "Middleware Services for Information Sharing in Mobile Ad-Hoc Networks - Challenges and Approach", Workshop on Challenges of Mobility, IFIP TC6 World Computer Congress, Toulouse, France, August 2004

24. Pužar, M., Plagemann, T., "NEMAN: A Network Emulator for Mobile Ad-Hoc Networks", Proceedings of the 8th International Conference on Telecommunications (ConTEL 2005), Zagreb, Croatia, June 2005

25. Rivest, R., "The MD5 Message-Digest Algorithm", RFC 1321, April 1992

26. Stajano, R., Anderson, R., "The Resurrecting Duckling: Security Issues for Ad-hoc Wireless Networks", 7th International Workshop on Security Protocols, Cambridge, UK, 1999

27. The OpenSSL project, http://www.openssl.org/

28. Tønnesen A., "Implementing and extending the Optimized Link State Routing protocol", http://www.olsr.org/, August 2004

29. Wallner, D., Harder, E., Agee, R., "Key management for Multicast: issues and architecture", RFC 2627, June 1999

30. Wong, C., Gouda, M. and S. Lam, "Secure Group Communications Using Key Graphs", Technical Report TR 97-23, Department of Computer Sciences, The University of Texas at Austin, November 1998

31. Zhou, L., Haas, Z., "Securing Ad Hoc networks", IEEE Network, 13(6):24-30, 1999

Remote Software-Based Attestation for Wireless Sensors

Mark Shaneck, Karthikeyan Mahadevan, Vishal Kher, and Yongdae Kim

Computer Science and Engineering,
University of Minnesota - Twin Cities

Abstract. Wireless sensor networks are envisioned to be deployed in mission-critical applications. Detecting a compromised sensor, whose memory contents have been tampered, is crucial in these settings, as the attacker can reprogram the sensor to act on his behalf. In the case of sensors, the task of verifying the integrity of memory contents is difficult as physical access to the sensors is often infeasible. In this paper, we propose a software-based approach to verify the integrity of the memory contents of the sensors over the network without requiring physical contact with the sensor. We describe the building blocks that can be used to build a program for attestation purposes, and build our attestation program based on these primitives. The success of our approach is not dependent on accurate measurements of the execution time of the attestation program. Further, we do not require any additional hardware support for performing remote attestation. Our attestation procedure is designed to detect even small memory changes and is designed to be resistant against modifications by the attacker.

1 Introduction

Recent technological advances in hardware and communications have helped to achieve significant strides in the area of wireless sensor networks. These networks can be used in several real-world applications, including various critical applications, such as military surveillance, infrastructure security monitoring and fault detection (e.g., Golden Gate Bridge monitoring [23]), or industrial waste monitoring.

When sensors are deployed for critical applications, securing these sensors is important. If a sensor is compromised, an attacker can reprogram the sensor to act on his/her behalf. For example, the attacker can cause the sensor to send incorrect information to hide some military activity or send false information about the location of certain troops. Therefore, it is important to verify that the static memory contents of the sensors have not been modified, that is, to *attest* the static memory contents (which includes programs, keys, and system configuration information) of the sensors. Typically, sensors are deployed in large numbers in environments that may not be safe or easily accessible to humans. Further, the deployment mechanisms (e.g., unmanned air planes) often make it infeasible to locate the position of each sensor individually. Therefore, we need

R. Molva, G. Tsudik, and D. Westhoff (Eds.): ESAS 2005, LNCS 3813, pp. 27–41, 2005.
© Springer-Verlag Berlin Heidelberg 2005

attestation mechanisms that do not require physical contact with the sensors, but rather use the wireless communication network. In other words, we need mechanisms to perform *remote* attestation.

In this paper we present a software-based approach to remotely attest the static memory contents of the sensors without requiring any additional hardware on the sensors. As sensors are inherently designed to be light-weight and inexpensive, adding additional hardware on the sensors significantly increases the cost well as the size of the sensors; therefore, software-based approaches are always preferable and practical (they also work on legacy systems). In our approach, in order to attest the sensor, the attester sends an attestation routine to the sensor and waits for some time (expected response time) to get a response from the routine. Once the response time elapses, the attester will not accept any response sent by the sensors. The sensor executes the routine, which randomly reads the sensor's static memory contents and returns a checksum of the memory contents. The attester has an exact image of the memory contents of each sensor and can pre-compute the checksum by locally running the attestation routine on the memory image. After receiving the checksum from the routine, the attester can verify whether the received checksum matches the expected result. Every attestation routine is unique per sensor and randomized so that the attacker will be unable to predict (and pre-compute the checksum) the next routine from the previous routines.

Motivation. One way of performing remote software-based attestation is to include a small attestation routine in the sensor's kernel that performs a checksum on the memory contents of the sensor. To prevent replay attacks, for every new attestation request, the attester sends a random key to the sensor and the routine on the sensor pseudo-randomly reads the memory contents and generates a checksum on these contents using the attester's key. However, this naïve approach is susceptible to a simple attack [32]. The attacker can modify the attestation routine such that instead of reading the sensor's memory contents, the routine reads the unmodified contents stored somewhere else by the attacker and computes a checksum on these contents. Since the routine is forced to read the unmodified memory contents, the checksum will be valid, and the attacker will be able to conceal his changes.

One important observation is that in this case, in order to generate a valid checksum, the attacker's modified attestation routine has to check before every memory read whether the current memory address belongs to the the modified portion of the memory by inserting if (or similar) statements. The attacker has to use static analysis techniques (that analyze program binary without executing it) to understand the routine and insert if statements within the routine, which also increases the execution time of the routine. The attester can use this fact to detect the attacker's modifications by measuring the actual time taken by the routine (running on the sensor) to generate the checksum and comparing it with the expected execution time. If the time taken to generate the checksum is greater than the expected time, the attester proclaims that the sensor is compromised. This approach was introduced by SWATT [32].

However, while performing software-based attestation over the network, the detection mechanism cannot be completely dependent on such minute execution delays, as the network and the current execution state of the sensor can introduce some unforeseen delays resulting into inaccurate measurement of the execution time of the attestation routine, and, thus, resulting in false positives or false negatives. Therefore, in order to accurately measure the execution delay, the attester should be in physical contact with the sensor, which cannot be always possible in practice.

Contributions. The main contributions of the paper are summarized as follows.

- We present an approach for detecting malicious changes to the static memory contents of wireless sensors. The approach allows the attester to attest the memory contents of the sensors over the network without requiring physical contact with the sensors.

- Our approach is not dependent on precise measurements of execution timing delays to detect malicious changes to the memory of the sensors. Therefore, our approach is more practical and can be used in real-world scenarios.

- Finally, the approach presented in this paper does not require any hardware support. Thus, we do not add any additional cost or increase the size of the sensor. Further, our approach can be easily applied on legacy systems.

Scope of this paper. This paper is focused on designing software-based attestation techniques that are secure against the static analysis attacks described above. We do not require any tamper-proof hardware on the sensors.

This paper is not focused on addressing the following impersonation attack, as detecting this attack requires additional tamper-proof hardware on the sensor. Consider an adversary that controls two identical sensors, or one sensor and one powerful machine that emulates the sensor. The attacker then modifies the memory contents of one sensor and keeps the other sensor or the emulated sensor unmodified. When the modified sensor receives an attestation routine, it forwards the routine to the other unmodified sensor or the emulated sensor on which it gets executed. Since the routine is executed as if on the original unmodified sensor, it will return a valid checksum and the modifications will go undetected. This attack can be detected by authenticating the actual processor that executed the authentication routine. However, this requires additional tamper-proof hardware on the sensor, e.g., controlled physical random functions [12, 13].

Organization. The remainder of this paper is organized as follows. Section 2 describes our system assumptions, requirements, and the attacker model. In section 3, we present the basic building blocks that are used to construct the attestation routine. Section 4 explains our attestation mechanism in detail. Section 5 presents security analysis of our system and an extension to our basic mechanism. Related work is presented in section 6 and section 7 draws conclusions and outlines future work.

2 Assumptions, Threat Model, and Requirements

2.1 Assumptions

The base station is assumed to be secure and it will play the role of an *attester* in our discussion. In reality, any legitimate entity that shares a pairwise key with the sensor can be an attester. The communications between the base station and the sensors is secure using a pairwise key shared between them. We do not address denial of service attacks (DoS) in this paper. The attester knows the hardware architecture and the original memory contents of the sensors. We assume that the sensors do not have virtual memory, as an attacker can modify the memory map, distinguish between data loads and instruction loads as pointed out in [11], and evade our attestation. We argue that this assumption is reasonable, since state of the art micro-controllers do not have virtual memory support [2, 3]. The attester can communicate with all the sensors directly. We also assume that the attester can send a binary executable to the sensor and cause it to be executed (e.g. [18]).

2.2 Threat Model

We assume that if the sensor is compromised, then the attacker has complete read-write access to the sensor's memory contents, including cryptographic keys, and is able to modify the memory contents at will. Thus, he can perform any type of software based attack on the attestation routine including static analysis (resulting in modification) of the routine, or software emulation of a sensor on a sensor. However, we assume that the attacker cannot tamper with the hardware of the sensor. Detection of attacks that involve external resources (such as the impersonation attack described in section 1) requires hardware support and is considered to be out of scope. We assume that the attacker can perform a re-stricted form of collusion attack, which we call as the staging attack. We assume that the attacker can execute the attestation routine in stages. For example, a sensor with some modified portion of the memory can collude with the second sensor with a different modified portion of the memory. Each sensor runs the routine in such a way that it generates checksum on their respective un-modified memory and then combine their checksums in the end to generate a valid check-sum. Finally, the attacker can perform passive attacks such as eavesdropping, and active attacks such as replaying packets.

2.3 Requirements

The attestation procedure should satisfy the following requirements.

- **Resistance to Replay**: The attacker should not able to send a valid check-sum to the verifier by simply replying previous valid results.
- **Resistance to Prediction**: The attacker should not be able to predict the next attestation routine. If the attacker can successfully predict the next attestation routine, then he can pre-compute the checksum.

- **Resistance to static analysis**: The attacker should not be able to success-fully analyze the code by using static analysis techniques within the time period the attester waits for a response from the sensor. This requirement will prevent the attacker from predicting the sequence of memory reads as well as predicting the location of read instructions in the attestation routine.
- **Very loose dependence on execution time**: Since the attestation routine is sent over the network, it will be impossible for the attester to measure the actual execution time of the attestation routine. Therefore, the detection mechanism should not be dependent on the precise measurement running time of the attestation routine.
- **Complete memory coverage**: To detect even small memory changes, the attestation routine should read every memory location.
- **Efficient construction**: The attestation routine should be as small as pos-sible to reduce bandwidth consumption and should be as efficient as possible to consume less battery power. Further, the attestation routine should not introduce any new vulnerability in the system.

3   Building Blocks

In order to prevent the attacks mentioned in Section 2.2, the attestation code will make use of the following building blocks. These constructs, which are described below, include randomization, encryption, obfuscation, and self-modifying code. These are not employed to provide unbreakable security, but rather they are used to make the aforementioned attacks infeasible to be carried out using a sensor's limited resources.

Randomization. The routine that is sent to the sensor to perform the attestation should be different each time. If the routine is different, in some random fashion, and the results of the attestation calculation are dependent on the specific version that is being run, then a previous version of the code could not be analyzed offline and reused later. Thus, the attacker is forced to perform the static analysis of the binary in an online fashion: the attacker needs to analyze and modify the routine and then execute it to return the result.

Encryption. The next construct that we use in the construction of the attestation code is encryption. We will make use of a simple encryption scheme (XOR each word with a random value) to prevent static analysis of the code directly. The attacker will thus need to first attack the decryption code in order to break the encryption of the remaining code. The encryption schemes are not meant to be secure in the traditional sense, but rather are aimed at adding complexity to the disassembly of the code. This technique has been explored previously in the field of software tamper resistance [4].

Self-Modifying Code. In addition, we use self-modifying code in the attestation program. Without this construct, an attacker could avoid doing the full static analysis of the code and just search for all memory read statements in the pro-gram. By doing this, the attacker can simply place conditional offsets before each read statement. However, if the reads are regenerated and rewritten in a different memory locations, then the attacker must first analyze the code that performs these writes. Without doing so the attacker could not reliably redirect the targets of these memory reads. The usage of this construct is explained further in Section 4.2 and has been proposed to strengthen operating system security [7].

Opaque Predicates and Pointer Aliasing. With the previous construct in place, the attacker is forced to analyze the entire program that it is sent. Thus we also add constructs to further complicate the task of static analysis as much as possible. For this purpose we use traditional obfuscation constructs, namely opaque predicates and pointer aliasing. Opaque predicates are predicates that always evaluate to either true or false, regardless of the input to the condition, yet it is very difficult to determine which branch will be taken each time, or even to determine whether this conditional is actually unconditional. Constructions of this type have been previously discussed in obfuscation literature [10, 8, 9]. One of the most promising constructions of opaque predicates is the use of pointer aliasing [28, 37] and performing data flow analysis of aliased pointers is known to be an NP-hard problem [17, 24, 30].

Junk Instructions. The use of junk or fake instructions can be combined with the opaque predicates described above to further confuse static analysis and disassembly [26]. Some of these junk instructions can be partial instructions, which will confuse the disassembly and thus hinder static analysis. Other instructions will be full instructions, which will be used to misdirect the static analysis and waste its time and efforts.

4   Design of Attestation Procedure

We now bring together all the building blocks described previously in Section 3 and describe our scheme to perform the attestation. Throughout the description of our scheme, we use the word code and routine interchangeably to refer to the attestation routine sent to the sensor.

4.1   Overview

The base station generates the attestation code, which will be sent to the sensor. The code construction is described in Section 4.2. When sending the code to the sensor, the base station encrypts the code and appends a MAC of the encrypted code, and sends this to the sensor. Upon receiving this message, the sensor first verifies the MAC and then decrypts the attestation code. The sensor then copies this into its program memory and transfers execution control to it. The attestation code will run and calculate the results. Once the result is calculated, it is sent back to the base station (again, this message is encrypted and authenticated by the sensor with the key it shares with the base station). Since the base station knows the image of the sensor's program memory, it can also run the code to compute the expected result. If the returned value matches the base station's expected result, then the sensor is declared to be ok. If the result is incorrect or if the sensor does not respond with the timeout period δ, then the sensor is declared to be corrupted. The base station should wait for a timeout period T_{wait} equal to $(2 * r) + e + \Delta$, where r is the time required to send a message from base station to the sensor (one way), e is the expected execution time of the attestation code, and Δ is a system parameter that indicates expected delay in the response due to network jitters, etc.

4.2   Attestation Code Construction

The high level construction of the code is as follows: the attestation code will generate random numbers (within the range of the sensor's memory that is to be attested) and reads the data at those memory locations. Those values will be hashed together incrementally (thus the order in which the data is read influences the final outcome of the attestation code). The code will also include each of the constructs mentioned in section 3, in order to prevent an attacker from modifying the code to avoid detection. This section will describe in detail how the code will be constructed to use those constructs.

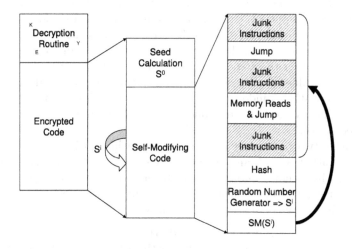

Fig. 1. Overall Structure

There are three main components in the code: the *seed calculation*, the *memory reads*, and the *hash computation*. For simplicity, we will describe the construction of the code with the assumption that each of these three parts are located in contiguous sections of the code and in the order described. However, the components can easily be interleaved with each other, with appropriate jumps between the different sections (obscured by opaque predicates). Figure 1 illustrates the construction of the attestation code.

Before discussing each component, we first describe the process of encryption. Not only is each component encrypted with a random key, but the entire attestation routine is also encrypted. Along with the encrypted code is the corresponding decryption routine. This will include the calculation of the key value, and the code to perform the decryption. The key will be located somewhere in the sensor's memory (either within the decryption routine itself, in some dead code space in the attestation routine, or using known portions of the sensors program memory), and so it will not be overly difficult for the attacker to discover the key. This discovery can be delayed, however, through the use of opaque predicates. Thus, we couple the key calculation with the opaque predicates, to obscure the location of the key (or components of the key). Then, the key will be calculated, which can be done in a number of ways, and the specific method used will vary randomly between each attestation routine. Some example mechanisms include taking the XOR of two (random) immediate values, adding together values from two "random" memory locations in the sensor's program memory, following multiple pointer indirections to the key value (where the pointers point to locations in the attestation routine). There are many such possibilities, and a few will be chosen at random (where one is the true method and its identity is hidden by the opaque predicates).

Seed Calculation. The first component of the code is the calculation of the seed, denoted as S^0 in Figure 1, which is used to initialize the pseudo-random number generator. As this value determines the order in which the memory contents are read, it will be in the attacker's best interest to leave this section of the code unmodified. Rather the attacker will need to determine a priori, through static analysis, the value of the seed. To prevent this, the seed calculation section is encrypted with a random encryption routine as described above. Also, the calculation of the seed will be done in the same manner as key calculation as described above.

The next two parts, memory reads and hashing, will be a part of a loop, where a location is read, and then the value is added to the hash computation. This loop will execute "enough" times to provide good coverage of the sensor's program memory (thus reducing the ability of the attacker to evade detection by hiding in a very small section of memory).

Memory Reads. The portion of the code that performs the memory reads is of particular importance, since that is where the attacker will attempt to inject the offsets in order to evade detection. This portion of the code has three main components: the *initial jump*, the *read* instruction, and the *self-modification*. The read instruction is initially located at some random address within this component, and so the jump instruction simply jumps control to this instruction. This jump, however, is obscured by the use of opaque predicates. In addition, as there is dead-code space, some of which will appear to the attacker to be reachable through the opaque predicate, junk instructions are inserted (randomly) in this space, both to thwart disassembly (with partial instructions) and to distract the static analysis (with normal instructions, such as memory reads). Following the read instruction, which places the contents of the particular memory address in question into a register, control is jumped to the self-modification section. This section is responsible for a number of tasks. First it generates three pseudo-random numbers. The first is used as the seed to the next iteration of the routine, denoted in Figure 1 as S^i. It uses the second as the next address to be read (and thus must be within the target range of addresses). The third is used to relocate the read instruction. It does this by overwriting the current read instruction with a junk instruction (or leaving it as is), and writing the new memory read instruction into another place in the code section. It also must update the initial jump so that control will be properly transferred to the new read instruction in the next iteration. This action is depicted in Figure 1 as $SM(S^i)$.

The random numbers will be generated using the RC4 pseudo-random number generator, as is used in SWATT [32]. In order to provide ample coverage of the memory space, it will iterate $O(n \log n)$ times though the memory read loop (this was shown in [32] to be a sufficient number of iterations to provide high coverage of the memory space), where n refers to the number of memory locations to be read.

Hash Computation. Next, the hashing component updates the current computation of the hash with the value that was read in the previous step. Once the computation of the hash is complete (all memory addresses are read), the final value is returned to the base station. We use the same hashing mechanism as in [32].

Construction by Base Station. The last item to be considered is the construction of each attestation routine by the base station. The base station must generate each attestation routine differently, such that the probability of two sensors receiving the same attestation code is very low (also the probability of a single sensor receiving the same code more than once should be very low). Thus each version of the attestation code must be generated randomly. This is achieved in several ways. First, the construction of the opaque predicates is based on the pointer-aliasing construction described in [9]. In our construction, these structures will be stored in random locations in the attestation code, and thus the opaque predicates will be different for each attestation routine. In addition, the seed will be chosen randomly, and the method used to compute this value will be chosen randomly from a set of possible methods.

5   Discussion

5.1   Security Analysis

In this paper we have presented a scheme for software based attestation that can be used in wireless sensor networks. In this section we will provide a discussion on the security properties of our scheme against the attacks described in Section 2.2.

First, an attacker can simply replay a previously computed response, or he could "sniff" a response from another sensor. However, in this case the attacker would only be successful if the seed used in the current attestation challenge is the same as the seed in the previous attestation challenge. Since the seed is chosen uniformly at random, this would only occur with negligible probability.

Thus the attacker must attempt to defeat the code contained in the current attestation challenge. Our goal is to force the attacker to perform some level of time intensive computation (which would delay the response to the attestation challenge past the timeout period). We argue that static analysis, while currently impossible to prevent, is computation intensive and will cause a significant delay in the response of the (resource-limited) sensor to the attestation challenge.

The attacker, then, has several options available in which to attack the code. First, the attacker can have an old version of the attestation code already analyzed (done offline at some previous point in time) and appropriately modified to avoid detection. In order for the attacker to be able to use this version of the attestation routine, he must first get the seed from the new attestation code. However, the attestation code, as well as the seed computation component, is encrypted. Thus the attacker must first break the two encryption schemes, which consists of determining the key that is used. This is protected by the opaque predicates. Also, once the attacker breaks the encryption schemes, he must determine the value of the seed, which is protected in the same way as the encryption keys. In order to accomplish these tasks, the attacker must perform static analysis on the code.

The attacker might also try to modify the read instructions, in order to insert the conditional offsets to redirect the read to the unmodified copy of the sensor's original code. The attacker could also use these modifications to redirect the reads to a collaborating sensor where that portion of the memory is unmodified (as per the staging attack described in Section 2.2). This also requires the attacker to determine the value of two encryption keys. In addition, due to the self-modifying code, the attacker cannot simply insert the conditional offsets into the code, but must first analyze the self-modifying portion of the code. By doing this, the attacker can cause the code to regenerate not only the memory reads, but the conditional offsets as well. Otherwise, if the attacker simply inserts the code before the initial read, the attestation code will overwrite this conditional offset and it will be lost. Thus, to do this, the attacker must again perform static analysis on the code.

Finally, the attacker can execute the attestation code within an emulator. In order for this attack to succeed, the attacker would pause execution of the code at each memory read, and offset the memory address to be read to point to an unmodified copy of the original sensor code. Emulation, however, imposes an inevitable slowdown in the execution of the program, and can be as much as an order of magnitude slower, as shown in [20]. Instructions are no longer decoded in hardware but in software. Also, code must be executed to process each emulated instruction. As this code is not bound by I/O, the slowdown will be significant, and with an appropriate choice of a timeout period, the base station can detect such an attack.

5.2   Extension

In addition, an optional extension to our scheme can be utilized to make emulation (and also static analysis) more difficult for the attacker to perform. During initial program code installation on the sensor, any free space in the sensor's program memory will be filled with random values. These random values will be known by the base station and thus can be included in the memory that is attested. Thus, the only free space available for the attacker to store an unmodified copy of the sensor's original code would be in the data memory. This is effective for two reasons. First, the data memory is typically used to store current execution information, such as the program stack, and thus certain portions of it cannot be overwritten by the attacker (without causing the sensor to crash). This not only reduces the amount of available space (thus requiring the malicious code to be very efficient), but also requires the attacker to be careful where the unmodified copy of the sensor's original code can be stored. Second, the data memory is typically much smaller than the program memory [1, 3], and thus the size of the malicious code (which performs static analysis or emulation) must be small enough to fit within the data memory.

6   Related Work

Software tamper-resistance is a technique to construct a program that either cannot be modified or an modification can be detected. There have been a variety of proposed approaches for achieving tamper-resistance. In general requiring additional hardware support has been one direction taken for solving this problem. On the other hand software based techniques such as obfuscation can be employed.

A trusted platform is one which adequately guarantees the users that the hardware and software modules are operating as specified. The Trusted Computing Group (TCG) has proposed an architecture called Trusted Platform Module (TPM). The TPM hardware, which is accompanied by supporting software, is used establish and provide a platform of trust. Load-time attestation using TPM is explored in [31]. BIND [34] employs TCG to perform fine-grained attestation; that is, it does not attest the entire memory but only a specific piece of code. Secure processors to prevent software tampering have been proposed in [25, 39, 35, 40]. Copilot [19] is a co-processor based runtime memory attestation mechanism. These hardware based approaches are not a suitable solution in our setting as sensors are expected to be inexpensive, and additional secure hardware would be prohibitively expensive.

SWATT [32], is a scheme that has been proposed to verify the static contents and configuration settings of an embedded device. However, as discussed previously in Section 1 this approach is not suitable for our setting. Genuinity [20], is a technique to ascertain whether a remote machine is running a real hardware running the expected software environment or not. Subsequently, attacks on Genuinity were described in [33, 32]. However in [21], the authors claim that the attacks on Genuinity are not sufficient to defeat the system. Our approach is similar in concept to Genuinity, in that we also send an attestation program to the sensor. As noted in [21], intricate details that could be exploited in embedded devices are rare, hence we have adopted sending a tamper-resistant attestation routine to achieve our goal.

The majority of the work on software based tamper-resistance relies on *ob-fuscation*. The goal of an obfuscating transformation is to make static analysis and disassembly of the executable, for the purpose of making useful modifications to a program, difficult [37, 10, 8, 9, 28, 38]. Theoretical work on obfuscation has yielded interesting results [14, 5, 36, 29, 27], and has shown that in general, perfect obfuscation is impossible. Therefore, careful choice of obfuscation transformations is necessary. For example, in [26], the authors proposed using indirect jumps (via branch functions) for preventing disassembly. By analyzing the control flow graph of the program and exploiting statistical techniques, the authors of [22] were able to correctly identify a majority of the program instructions.

Program evolution [7] was proposed as a technique to defend against automated attacks on operating systems. Self-checksumming software tamper resistance has been proposed in [6, 16]. Recently [11], the authors have shown the inadequacies of [6, 16] and proposed a generic attack on checksumming based software tamper resistance. The attack presented in [11] relies on advanced processor nuances like memory hierarchy, virtual memory and TLB, which is currently not available in sensors [1, 3], and hence is not applicable to our approach.

Integrity Verification Kernel (IVK) [4] is a technique for constructing tamper resistant *software*, where the software (that needs to be attested) is "armored" by means of self-encryption and self-decryption at run-time, coupled with self-checking of its integrity. This, however, is inherently different from our goals (attesting memory). If IVK is included in the sensor's programs, the attacker can simply reprogram the sensor. Further, the attacker can run the IVK in an emulator and get the actual (unencrypted) binary. If the attacker succeeds in getting the binary in clear, the attacker can generate a valid checksum on modified code. In our scheme, the attestation routine has to send a valid checksum on all of the static memory contents within the timeout period T_{wait}. Further, since the routine is new for each attestation, even if the attacker breaks one attestation routine, he cannot generate the checksum for the next attestation routine.

7   Conclusion and Future Work

Software attestation in sensor networks is one of the most important security primitives. To the best of our knowledge this effort is the first to consider remote software based attestation in sensor networks. We have presented a scheme which achieves this goal by sending a checksumming routine to the sensor from the base station. This code is protected by the techniques of encryption, obfuscation and self-modifying code, so that an attacker is unable to return a valid response from a compromised sensor within the allowed time. In addition, our approach is software based, and does not require the addition of any extra hardware.

Future work includes implementing and evaluating the presented attestation procedure. We are currently exploring ways to efficiently send the attestation routine to execute it on the sensor. We plan to explore the Mica Mote platform [2] and TinyOS [15], as this platform is known to support in-network reprogramming [18]. Detailed experiments will be performed to measure the overhead imposed by the attestation routine on the sensor in terms of battery consumption, code size, and execution time. Tests will also be performed on simulated sensors that can be used to simulate a large sensor network. As part of the experiments, we plan to measure the expected runtime so that we can provide estimates on the amount of time that the base station will wait for a response. We will also study the effect of Δ on the security of the system. Finally, a detailed security analysis of the implemented program will be provided.

References

1. Atmel AVR 8-bit RISC processor.
   http://www.atmel.com/atmel/products/prod23.htm.
2. Mica2 series.http://www.xbow.com/Products/Product_pdf_files/Wireless_pdf/
   MICA2_Datasheet.pdf.
3. TI MSP-430 processor. http://focus.ti.com/mcu/docs/techdocs.tsp?navSection=
   user_guides&templateId=5246&familyId=342.
4. D. Aucsmith. Tamper resistant software. In *Proceedings of the First Information Hiding Workshop*, 1996.
5. B. Barak, O. Goldreich, R. Impagliazzo, S. Rudich, A. Sahai, S. P. Vadhan, and K. Yang. On the (im)possibility of obfuscating programs. In *CRYPTO '01: Proceedings of the 21st Annual International Cryptology Conference on Advances in Cryptology*, pages 1–18, London, UK, 2001. Springer-Verlag.
6. H. Chang and M. J. Atallah. Protecting software code by guards. In *DRM '01: Revised Papers from the ACM CCS-8 Workshop on Security and Privacy in Digital Rights Management*, pages 160–175, London, UK, 2002. Springer-Verlag.
7. F. Cohen. Operating system protection through program evolution. Computers and Security, 1993.
8. C. Collberg, C. Thomborson, and D. Low. A taxonomy of obfuscating transformations. Technical report, Technical Report 148, Department of Computer Science, University of Auckland, July 1997.
9. C. Collberg, C. Thomborson, and D. Low. Manufacturing cheap, resilient, and stealthy opaque constructs. In *Principles of Programming Languages 1998, POPL'98*, San Diego, CA, Jan. 1998.
10. C. S. Collberg and C. Thomborson. Watermarking, tamper-proofing, and obfuscation - tools for software protection. In *IEEE Transactions on Software Engineering*, volume 28, pages 735–746, August 2002.
11. A. S. G. Wurster, P.C. van Oorschot. A generic attack on checksumming-based software tamper resistance. In *Proceedings of the IEEE Symposium on Security and Privacy*, May 2005.
12. B. Gassend, D. Clarke, M. van Dijk, and S. Devadas. Controlled Physical Random Functions. In *Proceedings of the 18th Annual Computer Security Conference*, December 2002.
13. B. L. P. Gassend. Physical random functions. Master's thesis, Massachusetts Institute of Technology, February 2003.
14. S. Hada. Zero-knowledge and code obfuscation. In *ASIACRYPT '00: Proceedings of the 6th International Conference on the Theory and Application of Cryptology and Information Security*, pages 443–457, London, UK, 2000. Springer-Verlag.

15. J. Hill, R. Szewczyk, A. Woo, S. Hollar, D. Culler, and K. Pister. System architecture directions for network sensors. In *ASPLOS-IX: Proceedings of the ninth international conference on Architectural support for programming languages and operating systems*, Cambridge, November 2000.

16. B. Horne, L. R. Matheson, C. Sheehan, and R. E. Tarjan. Dynamic self-checking techniques for improved tamper resistance. In *DRM '01: Revised Papers from the ACM CCS-8 Workshop on Security and Privacy in Digital Rights Management*, pages 141–159, London, UK, 2002. Springer-Verlag.

17. S. Horwitz. Precise flow-insensitive may-alias analysis is np-hard. *ACM Trans. Program. Lang. Syst.*, 19(1):1–6, 1997.

18. J. Jeong and D. Culler. Incremental network programming for wireless sensors. In *The First IEEE International Conference on Sensor and Ad hoc Communications and Networks*, October 2004.

19. N. L. P. Jr., T. Fraser, J. Molina, and W. A. Arbaugh. Copilot - a coprocessor-based kernel runtime integrity monitor. In *USENIX Security Symposium*, pages 179–194, 2004.

20. R. Kennell and L. H. Jamieson. Establishing the genuinity of remote computer systems. In *12th USENIX Security Symposium*, pages 295–310. USENIX Association, August 2003.

21. R. Kennell and L. H. Jamieson. An analysis of proposed attacks against genuinity tests. Technical report, Purdue University, 09 2004. CERIAS TR 2004-27.

22. C. Kruegel, W. Robertson, F. Valeur, and G. Vigna. Static disassembly of obfuscated binaries. In *Proceedings of USENIX Security 2004*, pages 255–270, San Diego, CA, August 2004.

23. T. Kuennen. Small science will bring big changes to roads. http://www.betterroads.com/articles/jul04a.htm.

24. W. Landi and B. G. Ryder. Pointer-induced aliasing: a problem taxonomy. In *POPL '91: Proceedings of the 18th ACM SIGPLAN-SIGACT symposium on Principles of programming languages*, pages 93–103. ACM Press, 1991.

25. D. Lie, C. Thekkath, M. Mitchell, P. Lincoln, D. Boneh, J. Mitchell, and M. Horowitz. Architectural support for copy and tamper resistant software. In *ASPLOS-IX: Proceedings of the ninth international conference on Architectural support for programming languages and operating systems*, pages 168–177, New York, NY, USA, 2000. ACM Press.

26. C. Linn and S. Debray. Obfuscation of executable code to improve resistance to static disassembly. In *CCS '03: Proceedings of the 10th ACM conference on Computer and communications security*, pages 290–299, New York, NY, USA, 2003. ACM Press.

27. B. Lynn, M. Prabhakaran, and A. Sahai. Positive results and techniques for obfuscation. In *EUROCRYPT '04*, 2004.

28. T. Ogiso, Y. Sakabe, M. Soshi, and A. Miyaji. Software tamper resistance based on the difficulty of interprocedural analysis, August 2002.

29. T. Ogiso, Y. Sakabe, M. Soshi, and A. Miyaji. Software obfuscation on a theoretical basis and its implementation. In *IEICE Transactions on Fundamentals*, volume E86-A, pages 176–186, January 2003.

30. G. Ramalingam. The undecidability of aliasing. *ACM Trans. Program. Lang. Syst.*, 16(5):1467–1471, 1994.

31. R. Sailer, T. Jaeger, X. Zhang, and L. van Doorn. Attestation-based policy enforcement for remote access. In *CCS '04: Proceedings of the 11th ACM conference on Computer and communications security*, pages 308–317, New York, NY, USA, 2004. ACM Press.

32. A. Seshadri, A. Perrig, L. van Doorn, and P. Khosla. SWATT: Software-based Attestation for Embedded Devicesi. In *Proceedings of the IEEE Symposium on Security and Privacy*, May 2004.

33. U. Shankar, M. Chew, and J. Tygar. Side effects are not sufficient to authenticate software. In *13th USENIX Security Symposium*. USENIX Association, August 2004.

34. E. Shi, A. Perrig, and L. V. Doorn. Bind: A time-of-use attestation service for secure distributed systems. In *Proceedings of the IEEE Symposium on Security and Privacy*, May 2005.

35. G. E. Suh, D. Clarke, B. Gassend, M. van Dijk, and S. Devadas. AEGIS: architecture for tamper-evident and tamper-resistant processing. In *ICS '03: Proceedings of the 17th annual international conference on Supercomputing*, pages 160–171, New York, NY, USA, 2003. ACM Press.

36. N. P. Varnovsky and V. A. Zakharov. On the possibility of provably secure obfuscating programs. In *Ershov Memorial Conference*, pages 91–102, 2003.

37. C. Wang, J. Hill, J. Knight, and J. Davidson. Software tamper resistance: Obstructing static analysis of programs. Technical report, University of Virginia, Charlottesville, VA, USA, 2000.

38. G. Wroblewski. *General Method of Program Code Obfuscation*. PhD thesis, Wroclaw University of Technology, Institute of Engineering Cybernetics, 2002.

39. J. Yang, Y. Zhang, and L. Gao. Fast secure processor for inhibiting software piracy and tampering. In *MICRO 36: Proceedings of the 36th Annual IEEE/ACM International Symposium on Microarchitecture*, page 351, Washington, DC, USA, 2003. IEEE Computer Society.

40. X. Zhuang, T. Zhang, and S. Pande. HIDE: an infrastructure for efficiently protecting information leakage on the address bus. In *ASPLOS-XI: Proceedings of the 11th international conference on Architectural support for programming languages and operating systems*, pages 72–84, New York, NY, USA, 2004. ACM Press.

Spontaneous Cooperation in Multi-domain Sensor Networks

Levente Buttyán, Tamás Holczer, and Péter Schaffer

Laboratory of Cryptography and System Security (CrySyS),
Department of Telecommunications,
Budapest University of Technology and Economics, Hungary
{buttyan, holczer, schaffer}@crysys.hu

Abstract. Sensor networks are large scale networks consisting of several nodes and some base stations. The nodes are monitoring the environment and send their measurement data towards the base stations possibly via multiple hops. Since the nodes are often battery powered, an important design criterion for sensor networks is the maximization of their lifetime. In this paper, we consider multi-domain sensor networks, by which we mean a set of sensor networks that co-exist at the same physical location but run by different authorities. In this setting, the lifetime of all networks can be increased if the nodes cooperate and also forward packets originating from foreign domains. There is a risk, however, that a selfish network takes advantage of the cooperativeness of the other networks and exploits them. We study this problem in a game theoretic setting, and show that, in most cases, there is a Nash equilibrium in the system, in which at least one of the strategies is cooperative, even without introducing any external incentives (e.g., payments).

1 Introduction

Multi-hop wireless sensor networks will be the near future's most powerful monitoring applications. These networks contain a large number of sensor nodes and some base stations which are collecting the information that the sensors measure. Sensor networks can be used for environmental monitoring (e.g., forest fire or earthquake detection), tracking of cars or material (e.g., freight transport, traffic monitoring), or monitoring the state of buildings [1].

An important design criterion for sensor networks is the minimization of the sensors' energy consumption. The reason is that sensors are often battery powered, and it is impractical, or in some cases, even impossible to change or recharge their batteries once they have been deployed. It is known that the energy consumption of transmitting a data packet is a super-linear function of the distance of the transmission. Practically, this means that, as far as energy consumption is concerned, it is more advantageous to transmit a packet in several small hops than to transmit it in a single large hop. Hence, if there are numerous sensors near to each other then they could transmit the packets together and by doing so, they can increase the lifetime of their batteries radically.

R. Molva, G. Tsudik, and D. Westhoff (Eds.): ESAS 2005, LNCS 3813, pp. 42–53, 2005.
© Springer-Verlag Berlin Heidelberg 2005

In today's research of sensor networks it is generally assumed that all the sensors and base stations belong to one authority that can control the whole network. In this paper, we depart from this common assumption, and consider sensor networks that are deployed at the same physical area, but controlled by different authorities. In such a situation, the sensors that belong to one authority may reduce their transmission energy even further if their packets are forwarded by sensors that belong to another authority; an act that we call *cooperation*. There is a risk, however, that the sensors belonging to the other authority are not willing to help and they drop the foreign packets.

We study this problem in a game theoretic setting. The main question we are interested in is the following: Can cooperation emerge spontaneously in multi-domain sensor networks based solely on the self-interest of the nodes (or more precisely the authorities to which the nodes belong)? To put it in another way: Is the objective of increasing the lifetime of the network enough to foster coopera-tion between co-located sensor networks? Our analytical and simulation studies presented in this paper show that in most cases, the answer to these questions is affirmative.

The rest of the paper is organized as follows. In Section 2, we show some cheering analytical results on a simplified model. In Section 3, we extend the simple model to a more realistic one, and we present our simulation results. In Section 4, we report on some related work. Finally, we conclude in Section 5.

2 Simplified Model

We start to study the problem of spontaneous cooperation in a simplified model. We assume that there are only two sensor networks that co-exist at the same physical location, and each of them consists of a single base station and a single sensor. The placement of the base stations and the sensors is illustrated in Figure 1.

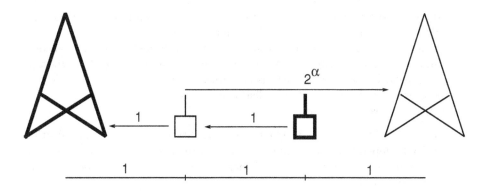

Fig. 1. Simple network

Now, we describe the operation of this simple system. We assume that time is divided into discrete time slots. In each time slot, each sensor wants to send a single data packet to its own base station, which contains its measurement data. We also assume that data packets are equal in size.

The packet can be sent to the base station directly in a single hop, or via the other sensor in two hops. Thus, at the beginning of each time slot every sensor has to decide the following:

- whether to request the other sensor to help in forwarding its own packet, or not, and
- whether to help in forwarding the other sensor's packet if it requests help, or not.

The decision made by the sensor defines its *move* in the time slot. Hence, we have four possible moves, each of which is denoted by a pair of letters as follows:

CC means that the sensor tries to get help from the other sensor and helps if the other sensor requests it;

CD means that the sensor tries to get help from the other sensor but it refuses to help if the other network requests it

DC means that the sensor does not ask for help, but it sends its packet directly to its base station, however, it helps if the other sensor requests forwarding;

DD means that the sensor does not ask for help from the other sensor, and does not help if the other sensor request forwarding.

In fact, C stands for cooperation and D stands for defection, and the first letter of the move defines how the sensor behaves concerning its own packet, and the second letter defines how it behaves when the other sensor's packet is concerned. For instance, making the move CD means that the node tries to cooperate when sending its own packet, but defects when the other sensor asks it to forward its packet.

Each pair of moves has a cost for both sensors, which are shown in Table 1. The costs are related to the energy consumption of the sensors and they are determined as follows:

- asking the other sensor to forward the packet has a unit cost, because this only requires to send the packet to a unit distance.
- forwarding the other sensor's packet also has a unit cost for similar reasons;
- sending the packet directly to the base station has a cost of 2^α, where α is the path loss exponent with usual values between 2 and 5, because this requires to send the packet to a distance of two units;
- dropping a packet has no cost.

We note that we are aware of the fact that in reality the cost of communication does not only depend on the distance, but there are also fix costs associated with the reception and the transmission of packets. In this simplified model, we set aside these fix costs.

The cells of Table 1 contain not only the costs for the two sensors, but also indicators of success, where 1 means that the packet reached the base station

Table 1. Costs and successes in the simple network (cost of row player, cost of column player; success of row player, success of column player)

	CC	CD	DC	DD
CC	$2,2$;$1,1$	$2,1$;$0,1$	$1,1+2^\alpha$;$1,1$	$1,2^\alpha$;$0,1$
CD	$1,2$;$1,0$	$1,1$;$0,0$	$1,1+2^\alpha$;$1,1$	$1,2^\alpha$;$0,1$
DC	$1+2^\alpha,1$;$1,1$	$1+2^\alpha,1$;$1,1$	$2^\alpha,2^\alpha$;$1,1$	$2^\alpha,2^\alpha$;$1,1$
DD	$2^\alpha,1$;$1,0$	$2^\alpha,1$;$1,0$	$2^\alpha,2^\alpha$;$1,1$	$2^\alpha,2^\alpha$;$1,1$

(success) and 0 means that it did not (failure). As an example let us consider the pair of moves $CC - CD$. In this case, the first sensor tries to send its packet via the other sensor, but the other sensor will drop it. On the other hand, the other sensor's packet will be sent via the first sensor to the base station successfully. Hence, the cost of the first sensor is 2 (1 for asking for forwarding its own packet and 1 for forwarding the other's packet), and the cost of the other sensor is 1 (the cost for asking for forwarding). Moreover, the first sensor records a failure, while the other one records a success.

We assume that the sensors record the results (success or failure) of the last few time slots in a buffer that we call *history*. One can think of the history as a binary vector of a fixed length. We assume that each sensor's next move is a function of its history. We call this function the *strategy* of the sensor.

Here we make an important restriction on the strategy space (set of possible strategies). We assume that each sensor wants to keep the weight of its history (i.e., the number of successful slots in the recent past) above a threshold, which we call the *weight threshold*. Intuitively, this means that we do not want to allow too many unsuccessful slots in the history, because that would mean that the base station does not receive measurement data with high enough rate (characteristic to the application). Therefore, when the weight of the history approaches the weight threshold, the sensor is not allowed to make risky $C*$ moves (i.e., CC and CD), but it is required to send its data directly to the base station (i.e., to make a $D*$ move). This situation is called the *constraint state*. Using strategies that suggest $D*$ moves in the constraint state guarantees that the weight threshold is never violated.

Note that a longer history with a lower weight threshold results in a system with more freedom. On the other hand, a shorter history with a higher threshold results in a much stricter system.

Each sensor has some initial battery B. In each time slot, the battery levels of the sensors decrease. The amount of this decrease depends on the pair of moves made in the time slot and their associated costs. When a sensor runs out of its battery, it dies. The other sensor can continue to send data to its base station if it still has some battery.

Note that the above mentioned concepts describe together an extensive game, where the players are the sensors, the possible moves (made simultaneously by both players) in each round (except for the constraint states) are CC, CD, DC, DD, the information sets are defined by the content of the histories, and the set of strategies are the functions that assign a move to every possible history with

Table 2. Best lifetimes with two-step strategies (lifetime for row player; lifetime for column player), B initial battery, ρ weight threshold, α path loss exponent, $\epsilon_{1,2}$ payoff from transient states

	CC/DD	CD/DD
CC/DD	$\frac{B}{2}$; $\frac{B}{2}$	$\frac{B}{\rho 2^\alpha+(1-\rho)}$; $\frac{B}{\rho 2^\alpha+(1-\rho)} + \epsilon_1$
CD/DD	$\frac{B}{\rho 2^\alpha+(1-\rho)} + \epsilon_1$; $\frac{B}{\rho 2^\alpha+(1-\rho)}$	$\frac{B}{\rho 2^\alpha+(1-\rho)} + \epsilon_2$; $\frac{B}{\rho 2^\alpha+(1-\rho)} + \epsilon_2$

the restriction that only $D*$ moves are assigned to a history that represents a constraint state. The game ends when both sensors run out of their batteries. The payoff of a player is its lifetime, which is represented by the number of rounds it survived. Lifetime is a good payoff function since the authorities want to run their network as long as possible with some constraints on their success.

Once we have a game, we can look for Nash equilibria with the highest possible lifetime. A Nash equilibrium is a strategy pair such that none of the players can increase its utility by unilaterally changing its strategy. It is quite reasonable to choose one of these Nash equilibria as an operating point in real systems. If there are more then one Nash equilibria, the equilibrium with the highest lifetimes is chosen.

In order to make the analysis feasible, we further restrict the strategy space. Let us consider first the *two-step strategies*. These strategies suggest a fix move if the player is not in a constrained state (independently of the actual weight of the history), and another fix move if the player is in a constrained state. A two-step strategy is denoted by m/m', where m is the move chosen in an unconstrained state and m' is the move chosen in a constrained state. For instance, the strategy CC/DD selects CC in an unconstrained state and DD in a constrained state. Therefore, we have eight two-step strategies, because in a constrained state only $D*$ moves are possible.

We performed an exhaustive search on this strategy space (there are $8 \times 8 = 64$ pairs of strategies to consider), and looked for Nash equilibria. We found that CC/DD and CD/DD dominate the other strategies. The CC/DD strategy is a cooperative strategy, while the CD/DD is an uncooperative one. By eliminating the dominated strategies, we get a reduced game. The lifetimes for the sensors in this reduced game are shown in Table 2, where ρ denotes the weight threshold, and B denotes the initial battery level. ϵ_1 and ϵ_2 comes from transient states like starting and ending the game. There are two Nash equilibria: $(CC/DD, CC/DD)$ and $(CD/DD, CD/DD)$. The first one results in full cooperation, while the second one results in full defection. However, if $\rho > \frac{1}{3}$ (and $\alpha \geq 2$ which is a fundamental condition in our model), then the cooperative equilibrium results in a higher lifetime for both players.

A more interesting class of strategies are the *weight aware strategies*. These strategies choose the next move as a function of the weight of the history. Thus, a weight aware strategy can be represented as $m_1/m_2/\ldots/m_k$, where m_1 is the move that is chosen when the weight of the history is maximal, and m_k is chosen when the weight of the history is just above the weight threshold. k is a parameter whose value depends on the history size and the value of the weight threshold. This class contains more complex and more reactive strategies.

After running 20 different exhaustive simulations, with different parameter sets, we found that the strategy that achieves the best Nash equilibrium is always the same: $(CD/CD/\ldots/CD/CC/DD)$. We call it the smart strategy. The smart strategy tries to ask for help in the first steps (the CD moves, which are cheap moves), but provides help (the CC move before the DD move) only in a state when the weight threshold is nearly violated in the hope that its nice behavior will be reciprocated. In other words, the smart strategy first tries to exploit the other. If this is successful, then it will never cooperate. However, if the other strategy is not exploitable, then it will change to a cooperative behavior. In the long run, the strategy keeps the actual weight of the history near to the weight threshold, which means that it cooperates only as much as necessary. This turns out to be a very effective behavior to save battery and leads to a rational cooperation.

In summary, we can see that in the simplified model, which contains two base stations and two sensor nodes, cooperative Nash equilibria exist based on smart strategies that try to optimize the amount of cooperation. In the next section we will investigate if the same is true in a more general model.

3   Generalized Model

After the cheering results of the simplified model in Section 2 we have examined much bigger and more complex systems. We have developed a simulator that corresponds to the model described in the first part of Section 2 with some extensions.

The generalized model uses many sensors per authority randomly placed on the playground with uniform distribution. The possible moves are the same as those in the simplified model, but in the generalized model each pair of moves has a cost that depends not only on the distance of the transmissions and the path loss exponent α, but also on some fix costs associated with the sending and receiving of packets. The fix cost of sending and the fix cost of receiving are constant values, which represent the energy consumption for connecting to the communication channel and to process the packets.

The principle of routing in the model is finding the minimum energy path towards the base station [9]. This means that every node has to forward on the path which has the minimum energy cost among all the possible paths. Every node maintains three paths: one in its own network (for the defective moves) and two in the global network (i.e., where all the nodes are possible forwarders). The global network paths are maintained for being able to make cooperative moves. The two distinct cooperative paths are towards the two base stations. These three paths can be the same depending on the placement.

Both networks have a threshold value (success threshold) which defines the minimum number of packets that the base station has to receive in each time slot, and the time slot is considered successful only if at least that number of packets reach the base station. The lifetime of a network is the total number

Table 3. Parameters for the simulations (the parameters are motivated in the *example* and in [7])

Parameter	Value
Number of sensors per domain	10-20-40 (20)
Distribution of the sensors	uniformly random
Area size	100x100 m
Position of the base (common base)	[50,50]
Position of the bases (separate bases)	[45,50] and [55,50]
Initial battery	10 million units
Reception fix cost	3000 units
Sending fix cost	2000 units
Success threshold	0.7-0.8-0.9 (0.8)
Weight threshold	0.6
History length	5
Energy drop-off (α)	2-3-4 (3)

of time slots that elapsed until the weight of the history becomes zero. The objective of the game is to reach the best possible lifetime under the constraint that the weight threshold of the history has to be respected.

Example: In an office building it is usual to deploy temperature and movement sensors. The temperature sensors measure the actual temperature and forward it to the air conditioning system. The movement sensors gather information about which zone is visited or abandoned and forward it to the security system. The two systems ask for information regularly (once in every second) but it is not crucial to get the information in every time slot. The temperature can be controlled and the security can be guaranteed with enough accuracy if some of the measurements are successful (let us say three out of the last five). The systems can work properly if they get enough measurement data in a time slot. While the sensors are usually deployed redundantly a given proportion can execute the task (let us say 80 % of the sensors). If the given proportion of data is arrived to the control systems, then the missing information can be deduced.

We have investigated two main type of scenarios. In one of them (common base scenario), there is a single common base station that collects the information from all of the nodes (independently from the authority they belong). In the other (separate base scenario), both networks have their own base stations. In the common base model, the base station is placed in the middle of the playground, while in the separate bases model, the base stations had the same distance from the theoretical middle of the playground.

We performed 100 simulation runs for each parameter setting with different topology. The concrete values for the simulations are shown in Table 3. The values in parenthesis are the defaults. For each run we made an exhaustive search in the strategy space to find the best strategy pairs (i.e., those that form a Nash equilibrium and generate the highest lifetimes).

In the extended model, it is not so easy to determine which equilibrium is a cooperative equilibrium. Two strategies can act in a cooperative way in case

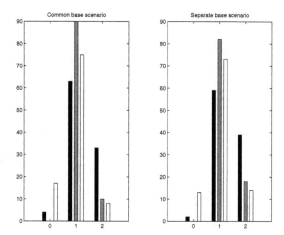

Fig. 2. Distribution of equilibrium classes (number of nodes per domain = 10 (black), 20 (gray), 40 (white))

of one topology and in an uncooperative way in case of another topology. In other words, the topology and the strategies both can influence the cooperation. Therefore, we classified the equilibria into the following three classes:

- *Class 0:* If the networks play strategies that form this type of equilibrium, then neither of them ever forwards a packet for the other. (no cooperation)
- *Class 1:* If the networks play strategies that form this type of equilibrium, then one of them forwards some packets for the other. (semi cooperation)
- *Class 2:* If the networks play strategies that form this type of equilibrium, then both of them forward some packets for the other. (full cooperation)

If the game had more than one best Nash equilibria, then we considered the most cooperative ones (i.e., those that have the highest class number).

The simulation results are shown in Figures 2, 3, and 4. In each figure, the left hand side chart shows the results of the common base scenario, and the right hand chart shows the results of the separate base scenario. On the x axis, we show the equilibrium classes (0, 1, 2), and on the y axis, the percentage of simulations where the best Nash equilibria fell in a given equilibrium class.

Figure 2 shows how the distribution of the different equilibrium classes depends on the number of nodes. One can see that in most cases the best Nash equilibria result in some kind of cooperation, although semi-cooperation has a higher probability than full cooperation.

Figure 3 shows how the distribution of the different equilibrium classes depends on the path loss exponent α. If α is high, then full cooperation is the best choice, because it costs a lot of battery energy to send to a far sensor. If full cooperation occurs, then the average sending distance is smaller, which is very advantageous when the path loss exponent is large.

Figure 4 shows how the distribution of the different equilibrium classes depends on the success threshold. One can see that the success threshold does not

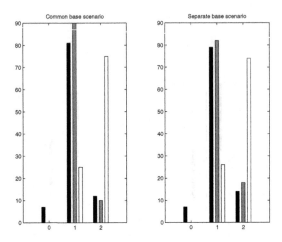

Fig. 3. Distribution of equilibrium classes ($\alpha = 2$ (black), 3 (gray), 4 (white))

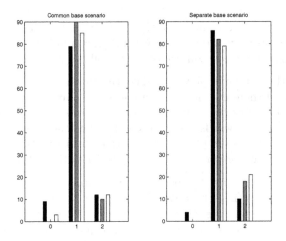

Fig. 4. Distribution of equilibrium classes (success threshold = 0.7 (black), 0.8 (gray), 0.9 (white))

have much influence on the distribution. If the success threshold is higher, than a little more fully cooperative Nash equilibria occur, but the success threshold seems to be a not as important parameter as the path loss exponent or the number of nodes.

As we have seen above, when co-located sensor networks are allowed to collaborate in the packet forwarding effort, some form of cooperation can emerge spontaneously, by which we mean that in the best Nash equilibria, at least one of the networks forwards some packets on behalf of the other network. It is clear that this cooperative behavior is more advantageous (meaning results in a longer lifetime) for the cooperating network than a defective behavior, given the

Table 4. Average gain in lifetime in the common base scenario and in the separate base scenario

Non-default parameter	Separate base scenario	Common base scenario
-	6.5%	6.1%
$n = 10$	15.5%	15.6%
$n = 40$	1.5%	0.6%
$\rho = 0.7$	4.4%	3.2%
$\rho = 0.9$	8.7%	7.8%
$\alpha = 2$	1.9%	2.2%
$\alpha = 4$	34.7%	31.0%

strategy of the other network, since a Nash equilibrium consists of best response strategies. In order to quantify this advantage, we performed the following experience. For each simulation run[1], we determined (i) the networks' lifetimes when both networks ignore each other and use only their own nodes for forwarding, and (ii) the networks' lifetimes in the best Nash equilibrium when the networks are allowed to collaborate. In both (i) and (ii), we took the smaller lifetime value (i.e., the lifetime of the network that lives shorter), and we computed the ratio of the values obtained. Finally, we averaged the ratio values over the 100 simulation runs (for each parameter setting). One can interpret the result of this computation as the average gain in lifetime when the networks are allowed to collaborate compared to the case when they operate independently from each other.

The results are shown in Table 4. Each row of the table belongs to a particular parameter setting, where all but one of the parameters have the default values shown in Table 3, and the first cell of the row shows the non-default parameter value. The second and the third columns of the table contain the average gain in lifetime in the common base and in the separate base scenarios, respectively. As we can see, the average gain in lifetime can be as high as 34% in the common base scenario and 31% in the separate base scenario when $\alpha = 4$.

4   Related Work

There are several articles that address the problem of cooperation in ad hoc networks (see e.g., [2, 3, 6]). However, these papers deal with the question of how cooperation can be encouraged by the introduction of some incentives (e.g., payments or reputations). Thus, indirectly, all these papers assume that cooperation cannot emerge by itself, but it must be stimulated. In contrast to this, we study spontaneous cooperation in this paper.

Cooperation without incentives has been studied in [8], but there are important differences between that paper and our work. First, the authors of [8] study cooperation in ad hoc networks, while we are considering cooperation in sensor

---

[1] Recall that for each parameter setting, we had 100 simulation runs with different topologies.

networks. Second, in [8], the nodes are collected into energy classes, which represent the heterogeneity of the nodes, whereas in our model, the nodes have equal resources. Finally, in [8], randomly chosen pairs of nodes communicate with each other, while in our case, every sensor communicates with the base stations.

In [4], the authors study the conditions under which cooperation (without incentives) is the best strategy in static ad hoc networks. Unlike the model in [8], their model takes into account the topology of the network. The main difference between [4] and our work is that energy consumption of the nodes is not considered in [4], whereas it has a central role in this paper. In addition, in [4], the nodes communicate with each other in a peer-to-peer manner, while we are considering sensors communicating with base stations.

The paper of Félegyházi et al. [5] stands most near to our work. In that paper, the authors investigated exactly the same problem as we do in this paper, nonetheless, their model and simulator is remarkably different from ours. First, in their model the payoff received after a successful round (in which enough sensors managed to send their data to the base station successfully) is a subjective value that represents the importance of a successful round for the given authority. In our case, there are no payoffs after the rounds, but instead the lifetime of the network is the payoff received at the end of the game. Second, in this paper, we introduce a constraint on the available moves after a certain number of unsuccessful rounds. This guarantees that a minimum level of quality of service is maintained in the network (i.e., base stations do receive data from sensors at least with a predefined rate), which indeed is a very important practical requirement. In [5], no lower bound on the success rate is guaranteed. Finally, we define the notion of lifetime differently: for us a network is dead when a certain percentage of its nodes die, while in [5], the death of the first node means the death of the whole network.

5   Conclusion and Future Work

In this paper we examined if cooperation is possible without the usage of incentive mechanisms in multi-domain sensor networks. First, we analyzed a simple network consisting of two sensors and two base stations, and found that in this simple setting, the best Nash equilibria (where the lifetime of the sensors is the highest) consist of cooperative strategies. Then we generalized our model from two nodes to many nodes, and used a two dimensional layout. We classified equilibria into non-cooperative, semi-cooperative, and fully cooperative. We found that in most cases, the best Nash equilibria belong to the cooperative classes. Especially, in the case when the path loss exponent is large, full cooperation is the best strategy.

In terms of future work, we intend to study more in detail how the distribution of the different equilibrium classes depend on the parameters the density and the topology. If this dependence can be characterized precisely, then it becomes possible to *engineer cooperation* by fine-tuning the parameters and adjusting the topology (if the application permits that) appropriately.

Acknowledgements

The authors are thankful to István Vajda for the helpful comments and discussions. The first author is also grateful to Jean-Pierre Hubaux and Márk Félegyházi for initial discussions on the simplified model presented in this paper.

This work has partially been supported by the Hungarian Scientific Research Fund (T046664). The first author has been further supported by IKMA and by the Hungarian Ministry of Education (BÖ2003/70).

References

1. I. F. Akyildiz, W. Su, Y. Sankarasubramaniam, and E. Cayirci. Wireless Sensor Networks: A Survey. *Computer Networks*, Vol. 38, No. 4, pp. 393-422, March 2002.
2. S. Buchegger and J-Y. Le Boudec. Performance Analysis of the CONFIDANT Protocol (Cooperation Of NodesFairness In Dynamic Ad-hoc NeTworks). In *Proceedings of the 3rd ACM International Symposium on Mobile Ad Hoc Networking and Computing (MobiHoc)*, pp. 80-91, June, 2002.
3. L. Buttyán and J.-P. Hubaux. Stimulating Cooperation in Self-Organizing Mobile Ad Hoc Networks. *ACM/Kluwer Mobile Networks and Applications (MONET)*, 8(5), October 2003.
4. M. Félegyházi, J.-P. Hubaux, and L. Buttyán. Nash Equilibria of Packet Forwarding Strategies in Wireless Ad Hoc Networks. *to appear in IEEE Transactions on Mobile Computing*
5. M. Félegyházi, J.-P. Hubaux, and L. Buttyán. Cooperative Packet Forwarding in Multi-Domain Sensor Networks. In *Proceedings of the First International Workshop on Sensor Networks and Systems for Pervasive Computing (PerSeNS)*, March 2005.
6. P. Michiardi, R. Molva. CORE: A COllaborative REputation mechanism to enforce node cooperation in Mobile Ad Hoc Networks. In *Communication and Multimedia Security 2002*, September 2002.
7. Rahul C. Shah, Jan M. Rabaey: Energy Aware Routing for Low Energy Ad Hoc Sensor Networks. In *IEEE Wireless Communications and Networking Conference (WCNC)*, 2002.
8. V. Srinivasan, P. Nuggehalli, C. F. Chiasserini, and R. R. Rao. Cooperation in Wireless Ad Hoc Networks. In *Proceedings of IEEE INFOCOM'03*, San Francisco, Mar 30 - Apr 3, 2003.
9. F. Ye, A. Chen, S. Lu, and L. Zhang. A scalable solution to minimum cost forwarding in large sensor networks. In *Proceedings of the Tenth International Conference on Computer Communications and Networks*, pp. 304-309, 2001.

Authenticated Queries in Sensor Networks

Zinaida Benenson

Department of Computer Science,
RWTH Aachen University, Germany

Abstract. This work-in-progress report investigates the problem of *authenticated querying* in sensor networks. Roughly, this means that whenever the sensor nodes process a query, they should be able to verify that the query was originated by a legitimate entity. I precisely define authenticated querying, analyze the design space for realizing it and propose solutions to this problem in presence of *node capture* attacks.

1 Introduction

Consider a sensor network which is deployed over a large geographic area. The maintainer of the sensor network offers services to the users: They can post queries to the sensor network using some mobile device. In this case, only the queries of legitimate users should be answered by the network. However, existing query processing systems (for an overview, see e.g. [26]) are not concerned with this issue. Meanwhile, this problem becomes especially difficult in presence of node capture attacks.

1.1 Node Capture in Sensor Networks

Node capture means gaining full control over a sensor through a physical attack, e.g., opening the sensor's cover and and reading out its memory and changing its program. A node capture attack can only be mounted on a small portion of the network if the network is sufficiently large, as direct physical access is needed.

This type of attack is fundamentally different from gaining control over a sensor remotely through some software bug, e.g., a buffer overflow. As all sensors are usually assumed to run the same software, in particular, the same operating system, finding an appropriate bug would allow the adversary to control the whole sensor network.

1.2 Motivation for Authenticated Querying: An Example

Directed Diffusion [12], a popular paradigm for organizing sensor networks, allows the user to post queries at any arbitrary sensor node (called the *sink*). The sink then floods the network with the query. After some time, sensor nodes start sending their aggregated data towards the sink. The sink gives the data to the user. In this case, to prevent the adversary from querying the sensor network, an access control mechanism should be built into each sensor node.

R. Molva, G. Tsudik, and D. Westhoff (Eds.): ESAS 2005, LNCS 3813, pp. 54–67, 2005.
© Springer-Verlag Berlin Heidelberg 2005

Consider an adversary who wants to gain unauthorized access to the data. He can try either to subvert the access control mechanism, or to find some weaker point in the sensor network architecture. For example, if the communication between the sensors happens without encryption and authentication, the adversary would bypass access control mechanism by directly attacking the communication protocol (eavesdrop, insert his own messages).

But even if all communication between the sensors is properly encrypted and authenticated, access control remains a separate problem which has to be solved. To illustrate this, consider a sensor network with Directed Diffusion mechanism where the sink is able to organize secure and authenticated communication with other sensor nodes.[1]

Suppose that, additionally to secure authenticated communication with the sink, some access control mechanism is built into each sensor node. This even could be an SSL/TLS-like protocol, as Gupta et al. [10] recently showed. Their implementation of SSL handshake on extremely resource constrained MICA2 sensors [5] takes less than 4 seconds.

However, if the adversary can disable the access control mechanism on a single sensor node, for example by capturing it, he would be able to query the entire sensor network. This single sensor, acting as a *new* sink, will build a secure authenticated channel to other sensor nodes, but this would not prevent the adversary from unauthorized data access. This happens because any arbitrary sensor is authorized to act on behalf of the user.

1.3   Contributions

The contributions are twofold:

1. This work systematically investigates the problem of authenticated queries in sensor networks in presence of node captures. To the best of my knowledge, such systematical approach was not considered previously. Moreover, solutions to certain problem instances in the literature are very scarce (see Section 7). A precise problem statement is given, and then the design space for solutions is specified.
2. For each possibility in the design space, solutions are discussed. Some of them are already known mechanisms for securing communication networks. For some other cases, original solutions are outlined. However, as this is a work-in-progress report, presented schemes are not fully implemented and analyzed yet.

Roadmap: Section 2 defines authenticated querying and gives design space for its implementation. Section 3 discusses general techniques for query authentication in sensor networks. In Sections 4 and 5, each possibility in design space is considered, and existing solutions are outlined. New solutions are presented in Section 6. Section 7 discusses related work. Section 8 summarizes and describes future work.

---

[1] I am not aware of such a mechanism for arbitrary sinks, although in [24], secure and authenticated variant of Directed Diffusion for a dedicated sink (the base station) is presented.

2   Problem Statement

2.1   System Model

Sensor network architecture. Consider a sensor network which is deployed over a large geographic area. The network consists of a large number of resource constrained sensor nodes such as MICA2 sensors [5] which have 128 KB flash instruction memory, 4 KB SRAM, an 8-bit microprocessor, and are powered by two AA batteries. The sensor nodes are not tamper proof, i. e., they are susceptible to node capture attacks.

There is also a small number of base stations which have more resources than the sensor nodes. For example, they can be laptop class devices.

Users. The maintainer of the sensor network offers services to a large number of mobile users. Legitimate users can access the sensor network using some mobile device like a PDA or a mobile phone.

Queries. Queries can be injected into the sensor network either at a base station (like in TinyDB [14] or Cougar [27]) or at any sensor node (like in Directed Diffusion [12]). The queries may be first optimized or otherwise processed at the place of injection and then they are disseminated in the sensor network using multihop communication according to some query processing mechanism.

Adversary. The goal of the adversary is to post arbitrary unauthorized queries to the sensor network. The adversary can capture a small amount of sensor nodes, read out their memory contents, and make them run arbitrary programs.

2.2   Authenticated Querying

As the adversary can capture some sensor nodes, he would have access to all data measured by these sensor nodes, and to all data routed through them (in case the data are unprotected or can be decrypted by means of captured cryptographic keys). This data disclosure cannot be prevented in face of node captures. Nevertheless, the adversary should not be able to post *arbitrary* queries to the sensor network.

Definition 1 (Authenticated Querying). *Let WSN be a sensor network consisting of N nodes $s_1, \ldots, s_N$. The users can post queries $q \in Q$ to the WSN. Consider an arbitrary query q. Let S_q be the set of all sensors which must process the query in order to give the required answer to the user. The WSN satisfies* authenticated querying *if it satisfies the following properties:*

- *(Safety) If a sensor s processes the query q, then q was posted by a legitimate user U.*
- *(Liveness) Any query q posted by a legitimate user U will be processed at least by all sensors $s \in S_q$.*

2.3 Design Space for Authenticated Querying

Two following dimensions can be identified for authenticated querying:

- The user has to communicate with the base station in order to post queries vs. the query can be started at some sensor nodes.
- The sensor network has to forward some data using multihop communication before the user can start posting queries vs. the query can be started locally.

These two dimensions give four possibilities for authenticated querying (AQ) (Table 1): direct base station AQ, remote base station AQ, distributed local AQ, and distributed remote AQ.

Table 1. Design space for realizing authenticated querying (AQ) in sensor networks. "Base station: yes" means that the base station must be accessed before the query processing can be started by the network. "Routing: yes" means that multihop communication is needed before the query processing can be started by the network.

	routing: no	routing: yes
base station: yes	direct base station AQ	routed base station AQ
base station: no	distributed local AQ	distributed remote AQ

Each of these mechanisms is appropriate in different situations and requires different solutions. In the following Section 3, general techniques for query authentication are discussed. In Sections 4 and 5, each mechanism from the design space is considered, and existing solutions are outlined. New solutions are proposed in Section 6.

3   Techniques for Authenticated Querying

Any of following techniques can be used with any mechanism from the design space. However, for clarity of presentation, I do not include them into Table 1, but list them separately.

3.1 Authenticated Broadcast

One possibility for an entity to authenticate its queries (or more generally, messages) is *authenticated broadcast*. This means that one sender can send a message to multiple receivers (here, the sensor nodes). The receivers can verify the origin of the message using some some authentication information attached to it.

Some approaches to authenticated broadcast in sensor networks exist in the literature. In SPINS [18], authenticated streaming broadcast μTESLA is realized using one-way hash chains, time synchronization, and symmetric keys shared by the base station with each sensor in the network.

Inexpensive digital signatures can also be used for authenticated broadcast (see e.g. [21]), assuming that each sensor node is preloaded with the public key of some certification authority. For discussion of public key cryptography in sensor networks, see Appendix A.

However, symmetric key cryptography should be preferred in sensor networks. Lower bounds on authenticated broadcast which uses only message authentication codes (MACs) are considered by Boneh et al. in [3]. Scheme for authenticated broadcast of Canetti et al. [4] meets this lower bound. It requires over 800 bits of authentication information per message, assuming that up to 10 receivers can collude and forge an authenticated message for a particular receiver with probability 2^{-10}. This scheme is independent from the number of receivers. Such high communication overhead is prohibitive in sensor networks. Below I argue that nevertheless, purely symmetric techniques can be considered for sensor networks.

3.2 Authenticated Flooding and Cooperative Approaches

Broadcast authentication protocol μTESLA achieves much better performance than the lower bound showed in [3]. This happens because, additionally to the system model used by Boneh et al., time synchronization is assumed between the sender and the receivers.

In sensor networks, another (implicit) assumption from [3] does not necessarily hold. This assumption states that the receivers do not communicate or cooperate with each other. However, in sensor networks, most queries are forwarded over multihop communication, or even flooded. Lower bounds from [3] do not apply if cooperation between the receivers is assumed. Therefore, more efficient methods which use symmetric cryptography may be possible. For an example of authenticated flooding, see Section 6.1.

An example of cooperative approach is interleaved message authentication from [22, 30]. There, sensor nodes on the multihop route along which a message is forwarded, cooperatively authenticate this message. This helps to withstand capture of some fixed number of sensor nodes. See Section 6.2 for an outline of an authenticated querying algorithm which uses interleaved message authentication.

4   Using Base Stations for Authenticated Querying

Base stations are supposed to have more resources and to be better protected against attacks than sensor nodes. Therefore, using base station is a natural approach to organize such a critical task as authenticated querying.

4.1 Direct Base Station Authenticated Querying

The query is always started at the base station, either by physically approaching it with a device and connecting to it (wirelessly or not), or by routing the query through some external network (e.g., the Internet) which is connected to the base station. Users log into the base station using an arbitrary client/server authentication protocol [16]. If the user is successfully authenticated, he can post arbitrary queries to the base station. The base station forwards (possibly optimized) user's queries into the sensor network.

In this case, the base station also has to authenticate its query such that all sensors which process the query can verify that it originated at the base station. However, the base station is a trusted entity, and therefore, has more opportunities for query authentication than a user. For instance, it may know the network topology, know at which nodes the data are stored, and share symmetric keys with each sensor in the network.

4.2   Remote Base Station Authenticated Querying

The query can be started on some sensor node (or nodes). Sensors are not concerned with query authentication, but just route the authentication information to the base station, the base station authenticates the user and then gives permission to the sensor network to answer user's queries. Thus, the base station helps to establish trust between the sensor network and the users. Here, at least two scenarios are possible.

(1) User's queries are always routed to the base station. The base station sends authenticated queries into the network on behalf of the user, receives the answers, and sends the answers back to the user. In this case, an SSL-like protocol can be used to set up a secure authenticated channel between the user and the base station.

(2) The base station generates a kind of "ticket" (using Kerberos terminology) which enables the user to talk to the sensor network for some time. The ticket is sent back to the user who uses it to generate authenticated queries. For a possible solution, see Section 6.1.

4.3   Authenticated Querying Using Base Station: Pros and Cons

(Advantages) The base station has more resources than a sensor and therefore, can implement stronger security measures. It can be placed in dedicated locations and maintained by humans. This makes the base station more reliable and secure: it can be protected from physical access, and DoS or penetration attacks are more likely to be spotted quickly.

(Disadvantages). The base station serves as a dedicated authentication server. Therefore, it must be very well protected from both physical and remote access by unauthorized entities. In the literature is usually assumed that to take over a base station is more difficult than a sensor node. In practice, the reverse could be the case. For example, if the base station is connected to a popular web server with known vulnerabilities, the penetration of the base station could be a matter of utilizing an available exploit. Besides, it is not always possible or desirable to place a base station into a dedicated secure location, especially if it is supposed to communicate with the sensors wirelessly.

Furthermore, if the direct physical access is needed for the authentication (e.g., wireless communication with the base station), it might be inconvenient to the user to walk through the half deployment area to the base station while needing the data from sensors in user's proximity. In case of remote access, several messages have to be routed between the base station and the user by the sensor network, which can be impractical if the base station is far away. The user might also have to wait for the answer of the far away base station while needing the data from sensors close-by.

And finally, as the sensors close to the base station are more heavily loaded with communication, their energy is exhausted more quickly, which leads to shorter network lifetime.

5   Authenticated Querying Without Base Stations

In cases where the base station cannot be used for authenticated queries, an access control mechanism or, more generally, a mechanism for query verification, should be built into sensor nodes. However, as shown in Section 1.2, relying on any arbitrary sensor for access control is not sound in face of node capture attacks. A natural solution here would be to use some kind of distributed algorithm, therefore the word "distributed" for the case without base station.

5.1   Distributed Local Authenticated Querying

The legitimacy of the query is verified by the sensors in user's location, e.g., in his communication range. Of course, even if the sensors in user's proximity successfully verified the query, they still need some means to tell to the rest of the sensor network that this query comes from a legitimate user. That is, at least sensors which process the query should be able to verify its legitimacy. For a concrete proposal, see Section 6.2.

5.2   Distributed Remote Authenticated Querying

The legitimacy of the query is verified by several sensors. These sensors can be specially chosen for this purpose, then the network architecture might be heterogeneous, with dedicated authentication devices placed in some locations. On the other hand, these sensors might be selected from the set of all sensors according to some algorithm.

Distributed remote authenticated querying can also be organized if each sensor in the network can verify the legitimacy of the query. For example, each user could receive a private/public key pair, certified by the certification authority of the maintainer, for signing his queries. Each sensor in this case must be preloaded with the authentic copy of the public key of the certification authority. Each query is then digitally signed and sent into the sensor network together with the corresponding user's certificate. In order to reduce computational burden on the sensor nodes, each sensor node could decide whether it is to verify the query with some probability.

According to Appendix A, signature verification can be an efficient operation. However, overhead for verifying user-generated signatures is twice as large as for signatures generated by the base station (see Appendix A.3).

5.3   Authenticated Querying Without Base Station: Pros and Cons

(Advantages) Users can start queries locally in the sensor network, without going to the base station or some other access point (e.g., an Internet terminal). Furthermore, the query can be processed and answered locally. No routing to the base station is needed for answering the query. Still, routing can be needed if the query concerns sensor data which are not in the user's proximity. And last but not least, if the base station is overloaded or taken over, the data are still available and not compromised (at least, as long as the compromised base station can be excluded from the network management).

(Disadvantages) The most severe disadvantage is that user authentication costs extra computation and communication power, especially if it is done in distributed fashion, using replication and agreement techniques, or public key cryptography. Distributed algorithms have to be applied in order to cope with unreliability and insecurity of individual nodes.

6   New Ideas for Authenticated Querying

Here, I present new proposals and ongoing work on authenticated querying.

6.1   Ticket Generation for Remote Base Station Authenticated Querying

I propose ticket generation using ID-based key predistribution. Key predistribution for sensor networks originates from [8]. The idea is that each sensor node is preloaded with randomly chosen m keys form the key pool of size l. The values of l and m can be chosen such that any two nodes have at least one common key with a given probability. ID-based key predistribution was introduced in [31]. The keys in the key pool are numbered from 1 to l. Each sensor with a unique identifier id is first assigned m distinct integers between 1 and l by applying a pseudo random number generator $PRG()$ with seed id. This method of choosing keys enables any sensor node u which knows the identifier id_v of another sensor node v to compute v's key identifiers by computing $PRG(id_v)$ and thus determine if u and v share some keys.

I adapt the above scheme to authenticated querying. In my idea, keys (and their identifiers) are predistributed to the sensor nodes using $PRG(id_{sec})$ where id_{sec} are *secret* sensor identifiers known only to the base station, but not stored on the sensor nodes. These identifiers should be sufficiently large (e.g., 80 bits) to prohibit the brute-force search of an identifier which generated a particular set of key identifiers. Therefore, an adversary who captured a sensor node cannot determine the secret identifier id_{sec} from which key identifiers were derived.

If a user U successfully authenticated to the base station, he receives from the base station a temporal identifier id_U (with a time stamp) and a set K_U of m secret keys which the base station computed using $PRG(id_U)$. Now the user formulates his query q and computes $h(q)$ where h is a hash function. After that, the user computes m 1-bit message authentication codes (MACs) on $h(q)$ using keys from K_U. The idea of using MACs with single bit output originates from [4].

The query accompanied by m 1-bit MACs and user's temporal identifier id_U is sent into the sensor network. Each sensor s can compute $PRG(id_U)$ and thus determine whether some of 1-bit MACs were computed with keys known to it. Then s verifies all 1-bit MACs which it is able to verify. If any of them is wrong, the node discards the query. Otherwise, it forwards the query to its neighbors.

This scheme is an example of authenticated flooding (see Section 3.2). The query of a legitimate user will be flooded into the sensor network without any obstacles. However, a query forged by an adversary will only be able to reach a limited part of the network, as some sensor nodes will discard the query. An initial analysis of query propagation in this model shows that a relatively small number of 1-bit MACs suffices to limit the query propagation to a logarithmically small part of the network. Thus, if there are N sensor nodes, approximately $\ln N$ sensor nodes will receive a fake query. For example, for $N = 10,000$, 300 1-bit MACs suffice. Further analysis of this scheme is subject to ongoing work.

The communication overhead of this scheme is rather high. If we assume that each query is accompanied by 300 bits of MACs and a user identifier which is 80 bits long, this results in 380 bits, or 48 bytes, of authentication information. The payload of a TinyOS[2] message is 29 bytes. Thus, each query is accompanied by two packets of authentication information. However, in contrast, an RSA-1024 digital signature which provides very strong message authentication (each sensor can verify the legitimacy of each message) is 1024 bits long, and therefore, requires five TinyOS packets. See Appendixes A.2 and A.3 for more information about using digital signatures in sensor networks.

6.2   Distributed Local AQ

If the number of users is large, the natural method to use for authentication is public key cryptography because of its scalability. On the other hand, public key cryptography is power-hungry, so the sensors should communicate with each other using symmetric cryptography. My concept is to let the sensors in the communication range of the user serve as interpreter (or a gateway) between the "public key crypto world" of the user and the "symmetric crypto world" of the WSN. The user talks to sensors in his communication range using public key cryptography, and these sensors then talk to the remainder of the sensor network on behalf of the user using symmetric cryptography. This happens in authenticated fashion:

---

[2] TinyOS is a popular operating system for sensor networks, see e.g. [13].

1. *Robust secure channel setup between the user and the WSN*: The user executes a mutually authenticated key establishment protocol [16] using public key cryptography with a specified number m of sensors in his communication range. The protocol results in establishment of symmetric session keys between the user and each correct sensor which participated in a protocol run.

2. *Authenticated query forwarding*: After the successful secure channel setup, the sensors in user's proximity forward user's query into the sensor network and append to it some additional information which enables the other sensors to verify the legitimacy of the query. In this case, the other sensors should be able to verify that at least m sensors approved the query.

We partially implemented the first step on Telos Revision B sensor nodes [17] using the EccM library for elliptic curve cryptography (ECC) [15]. For details, see [1]. In our implementation, the user unilaterally authenticates to the sensors in his proximity using public key cryptography. Moreover, a great breakthrough was recently reported by Gupta et al. [10]. They implemented the SSL protocol on MICA2 motes using elliptic curve cryptography. This confirms the choice of ECC for our implementation and the feasibility of mutually authenticated key establishment on sensor nodes, as their SSL handshake takes less than 4 seconds.

One possible solution to the second step (query forwarding) would be to use interleaved message authentication [22, 30] with $m - 1$ as a parameter for the number of captured sensor nodes. This is our ongoing work.

7   Related Work

In SPINS [18], authenticated streaming broadcast μTESLA is realized using one-way hash chains and time synchronization. In [24], one-way hash chains help to authenticate queries in Directed Diffusion. Inexpensive digital signatures are used in [21].

LEAP [29] and LKHW [19] consider using a single symmetric key for authenticated querying and therefore, are not resistant against impersonation attacks which are possible in case of node captures. Hierarchical sensor network architecture is considered in [6], [7]. This helps to support in-network processing, but the cluster heads become very attractive targets for node capture attacks.

All above approaches are suitable for sensor networks with the large number of users only in case of direct base station authenticated querying. To the best of my knowledge, the problem of enabling a large number of mobile users to post authenticated queries to a sensor network was not considered previously.

There is also a number of papers where the "opposite direction" is considered, e.g., verification of the legitimacy and correctness of answers to the queries. These methods cannot be directly applied to authenticated querying, as they assume that the verifier (the base station) has more resources than the sensor nodes. Statistical approaches are considered in [20,23,28]. Using clustered sensor network architecture and symmetric key techniques is described in [21,30].

8   Conclusions and Future Work

I defined the problem of *authenticated querying*, systematically examined the design space for its realizing, considered existing solutions and proposed some new methods for authenticated querying. These are first results of an ongoing project on access control to sensor network data. There is still a lot of work to be done. Proposed solutions have to be precisely specified. Security and resource demands of all solutions have to be more detailed analyzed, theoretically as well as by simulations and by implementation on real sensor nodes. First steps towards these goals are made in [2, 1].

References

1. Z. Benenson, N. Gedicke, and O. Raivio. Realizing robust user authentication in sensor networks. In *Real-World Wireless Sensor Networks (REALWSN)*, Stockholm, June 2005.

2. Z. Benenson, F. C. Gärtner, and D. Kesdogan. An algorithmic framework for robust access control in wireless sensor networks. In *Second European Workshop on Wireless Sensor Networks (EWSN)*, January 2005.

3. D. Boneh, G. Durfee, and M. K. Franklin. Lower bounds for multicast message authentication. In *Proceedings of the International Conference on the Theory and Application of Cryptographic Techniques*, pages 437–452. Springer-Verlag, 2001.

4. R. Canetti, J. Garay, G. Itkis, D. Micciancio, M. Naor, and B. Pinkas. Multicast security: A taxonomy and some efficient constructions. In *Proc. IEEE INFOCOM'99*, volume 2, pages 708–716, New York, NY, Mar. 1999. IEEE.

5. Crossbow, Inc. MICA2 data sheet. Available at
   http://www.xbow.com/Products/Product_pdf_files/
   Wireless_pdf/MICA2n_Datasheet.pdf.

6. J. Deng, R. Han, and S. Mishra. Security support for in-network processing in wireless sensor networks. In *SASN '03: Proceedings of the 1st ACM workshop on Security of ad hoc and sensor networks*, pages 83–93, New York, NY, USA, 2003. ACM Press.

7. T. Dimitriou and D. Foteinakis. Secure and efficient in-network processing for sensor networks. In *First Workshop on Broadband Advanced Sensor Networks (BaseNets)*, 2004.

8. L. Eschenauer and V. D. Gligor. A key-management scheme for distributed sensor networks. In *Proceedings of the 9th ACM conference on Computer and communications security*, pages 41–47. ACM Press, 2002.

9. G. Gaubatz, J.-P. Kaps, and B. Sunar. Public key cryptography in sensor networks - revisited. In *ESAS*, pages 2–18, 2004.

10. V. Gupta, M. Millard, S. Fung, Y. Zhu, N. Gura, H. Eberle, and S. C. Shantz. Sizzle: A standards-based end-to-end security architecture for the embedded internet. In *Third IEEE International Conference on Pervasive Computing and Communication (PerCom 2005)*, Kauai, March 2005.

11. N. Gura, A. Patel, A. Wander, H. Eberle, and S. C. Shantz. Comparing Elliptic Curve Cryptography and RSA on 8-bit CPUs. In *CHES2004*, volume 3156 of *LNCS*, 2004.

12. C. Intanagonwiwat, R. Govindan, D. Estrin, J. Heidemann, and F. Silva. Directed Diffusion for wireless sensor networking. *IEEE/ACM Trans. Netw.*, 11(1):2–16, 2003.
13. P. Levis, S. Madden, D. Gay, J. Polastre, R. Szewczyk, A. Woo, E. Brewer, and D. Culler. The emergence of networking abstractions and techniques in tinyos. In *First USENIX/ACM Symposium on Networked Systems Design and Implementation*, 2004.
14. S. Madden, M. J. Franklin, J. M. Hellerstein, and W. Hong. The design of an acquisitional query processor for sensor networks. In *SIGMOD '03: Proceedings of the 2003 ACM SIGMOD international conference on Management of data*, pages 491–502, New York, NY, USA, 2003. ACM Press.
15. D. J. Malan, M. Welsh, and M. D. Smith. A public-key infrastructure for key distribution in TinyOS based on elliptic curve cryptography. In *First IEEE International Conference on Sensor and Ad Hoc Communications and Network*, Santa Clara, California, October 2004.
16. A. J. Menezes, P. C. V. Oorschot, and S. A. Vanstone. *Handbook of Applied Cryptography*. CRC Press, Boca Raton, FL, 1997.
17. Moteiv, Inc. Tmote Sky datasheet. Available at http://www.moteiv.com/products/docs/tmote-sky-datasheet.pdf.
18. A. Perrig, R. Szewczyk, V. Wen, D. Culler, and J. D. Tygar. SPINS: security protocols for sensor netowrks. In *Proceedings of the 7th annual international conference on Mobile computing and networking*, pages 189–199. ACM Press, 2001.
19. R. D. Pietro, L. V. Mancini, Y. W. Law, S. Etalle, and P. J. M. Havinga. LKHW: A Directed Diffusion-Based Secure Multicast Scheme for Wireless Sensor Networks. In *32nd International Conference on Parallel Processing Workshops (ICPP 2003 Workshops)*, 2003.
20. B. Przydatek, D. Song, and A. Perrig. SIA: Secure information aggregation in sensor networks. In *ACM SenSys 2003*, Nov 2003.
21. S. Seys and B. Preneel. Efficient cooperative signatures: A novel authentication scheme for sensor networks. In *2nd International Conference on Security in Pervasive Computing*, number 3450 in LNCS, pages 86 – 100, April 2005.
22. H. Vogt. Exploring message authentication in sensor networks. In *Security in Ad-hoc and Sensor Networks (ESAS), First European Workshop*, volume 3313 of *Lecture Notes in Computer Science*, pages 19–30. Springer, 2004.
23. D. Wagner. Resilient aggregation in sensor networks. In *SASN '04: Proceedings of the 2nd ACM workshop on Security of ad hoc and sensor networks*, pages 78–87. ACM Press, 2004.
24. X. Wang, L. Yang, and K. Chen. Sdd: Secure distributed diffusion protocol for sensor networks. In *First European Workshop on Security in Ad-hoc and Sensor Networks(ESAS)*, volume 3313 of *Lecture Notes in Computer Science*, pages 205–214, 2004.
25. R. Watro, D. Kong, S. fen Cuti, C. Gardiner, C. Lynn, and P. Kruus. TinyPK: securing sensor networks with public key technology. In *Proceedings of the 2nd ACM workshop on Security of ad hoc and sensor networks*, pages 59–64. ACM Press, 2004.
26. A. Woo, S. Madden, and R. Govindan. Networking support for query processing in sensor networks. *Commun. ACM*, 47(6):47–52, 2004.
27. Y. Yao and J. Gehrke. The cougar approach to in-network query processing in sensor networks. *SIGMOD Rec.*, 31(3):9–18, 2002.
28. F. Ye, H. Luo, S. Lu, and L. Zhang. Statistical en-route detection and filtering of injected false data in sensor networks. In *Proceedings of IEEE INFOCOM*, 2004.

29. S. Zhu, S. Setia, and S. Jajodia. LEAP: efficient security mechanisms for large-scale distributed sensor networks. In *Proceedings of the 10th ACM conference on Computer and communication security*, pages 62–72. ACM Press, 2003.

30. S. Zhu, S. Setia, S. Jajodia, and P. Ning. An interleaved hop-by-hop authentication scheme for filtering of injected false data in sensor networks. In *IEEE Symposium on Security and Privacy*, pages 259–271, 2004.

31. S. Zhu, S. Xu, S. Setia, and S. Jajodia. Establishing pair-wise keys for secure communication in ad hoc networks: A probabilistic approach. In *IEEE International Conference on Network Protocols*, November 2003.

A   Asymmetric Key Cryptography in Sensor Networks

A.1   Efficiency Considerations

Although symmetric key cryptography is several orders of magnitude more efficient than asymmetric key cryptography [16], quite a number of researchers considered implementing efficient public key cryptography on small devices in the last few years. The reasons are that asymmetric key cryptosystems scale much better and allow more flexible key management, e.g., key agreement. Moreover, some asymmetric key cryptosystems allow efficient algorithms for encryption and for verification of digital signatures. In contrast, digital signature schemes based on symmetric mechanisms usually require large verification keys (cf. [16, page 31]).

A.2   Choosing Appropriate Asymmetric Key Cryptosystem

RSA with small public exponent (considered in [11, 25]) and Rabin public key cryptosystems (considered in [9]) have fast algorithms for encryption and digital signature verification. However, decryption and signature generation are slow and resource-demanding. Therefore, these cryptosystems can be used in sensor networks only if the sensors are not required to decrypt or to sign messages.

In contrast, elliptic curve cryptosystems (ECC, considered in [15]) require more overhead for encryption and signature verification than for decryption and signing. Nevertheless, with ECC, not only encryption and signature verification, but also decryption and signing are feasible for sensor nodes. Recently, Gupta et al. [10] implemented a very efficient SSL-like protocol on MICA2 sensor nodes using ECC in assembly language.

A.3   Digital Signatures by Base Stations vs. by Users

In some algorithms discussed in the main part of this paper, sensor nodes have to verify digitally signed messages originated from the base station or from the users. Each sensor has an authentic copy of base station's public key preloaded. To verify signatures of the base station, a sensor node just needs to apply the preloaded public key of the base station to the signature.

To be able send digitally signed messages to the sensor nodes, each user receives from the base station a certificate. This certificate is essentially user's public key signed by the base station. To verify user's signature of a message, the sensor network needs first to verify user's certificate. Thus, in case of user-signed messages, a sensor node needs to verify two signatures: First, the base station's signature on the certificate, and second, user's signature on the message. This means that verifying user-generated signatures roughly requires twice more resources than verifying signatures generated by the base station.

Improving Sensor Network Security with Information Quality

Qiang Qiu, Tieyan Li, and Jit Biswas

Institute for Infocomm Research (I^2R),
21 Heng Mui Keng Terrace, Singapore 119613
{qiu, litieyan, biswas}@i2r.a-star.edu.sg

Abstract. With extremely limited resources, it is hard to protect sensor networks well with conventional security mechanisms. We study a class of passive fingerprinting techniques and propose an innovative *information quality* based approach to improve the security of sensor network. For each sensor, we create a quality profile QP of profiling its normal/standard sensing behaviour. After deployment, new sensor readings are verified using this QP. If significant deviation is found, we either regard the readings as an abnormal behaviour or declare the sensor to be a fake sensor. The methods can be used as an assistant sensor authentication mechanism, but with a potential drawback. Furthermore, we also demonstrate a secure data fusion protocol, applying the proposed methods together with conventional security mechanisms. Through security analysis, we point out several countermeasures that can explicitly or implicitly defend against these attacks.

1 Introduction

Sensor networks are increasingly important for a wide variety of applications such as health monitoring, environmental control and military surveillance. With cheaper and smaller sensors, they are more attractive to be deployed in large scale multi-purpose applications in different scenarios. Security is an important issue [1] when sensors are used in military applications or in safety-critical applications, e.g. medical monitoring. However, sensors are designed with primary goals of smallness, cheapness and power-saving, rather than security. On the other hand, wireless sensor networks, communicating over open medium, are more vulnerable to passive/active attacks. Conventional security mechanisms, applied widely for protecting PCs from Internet attacks, are not effective in protecting sensors under the assumptions of extremely constrained resource and more powerful adversaries (who are able to launch arbitrary attacks). For instance, one can not differentiate a sensor reading taken on a genuine sensor from one taken on a fabricated sensor, since both readings are encrypted/authenticated by the same (fabricated) key.

Authentication of a sensor reading is very challenging under this extreme condition. Traditional authentication means may check what a sensor knows (e.g. a secret key), what a sensor has (e.g. a security token) or what a sensor is

R. Molva, G. Tsudik, and D. Westhoff (Eds.): ESAS 2005, LNCS 3813, pp. 68–79, 2005.
© Springer-Verlag Berlin Heidelberg 2005

(e.g. a fingerprint). Node fabrication attack makes the first two attempts fail. The third one-fingerprinting is not new and adopted by many practical systems (e.g. biometric authentication system, remote OS fingerprinting on Internet, etc.). We study the fingerprinting technique to be used in the context of sensor, that is *"Can we distinguish a sensor from others by investigating some unique physical behavior of a sensor?"*. In other words, we compare the new readings with some pre-sampled patterns of the same sensor and look for a match to determine the origin of these readings. Our approach assumes a "passive fingerprinting" situation, since the readings are typically received passively (not interactively or actively) from a sensor. Furthermore, since the sensor reading has to be concise to reduce power consumption on transmission, it does not contain additional informative data like a time stamp, the justifiable evidence only consists of a series of fresh sensor measurements. Thus, our method could be convincing if it is able to make accurate decisions depending on these raw measurement data.

In this paper, we introduce *information quality* for these measurement data. In our general model, the sensor nodes are calibrated statistically with its information quality profile (QP) before deployment. The profiles are then used as the references for sensors' reinforcement and rectification. Through experiments, we test how sensor quality is affected by attacks, how these sensors are detected and how the sensor network is protected. The success of the proposed approach may provide another level of security for sensor networks. We can foresee that it is applicable in many security critical applications. Our contributions are: 1, A novel approach of using information quality for improving the security of sensor network. 2, We demonstrate a secure data fusion protocol of embedding our methods with traditional security mechanisms. 3, We analyze the attacks on sensor network and propose our countermeasures using QP.

2   An Approach Using Information Quality Profile

Sensor networks are normally designed to minimize the sensors' power consumption, in most cases, only the measurement data are transmitted to the base station. Other informative (and relevant) data, such as time stamp, is not included. Our method utilizes the measurement data only and draws the profile for sensors. This section introduces how to generate quality profile (QP) and how to use it to detect sensors' misbehaviours.

2.1   QP and IQP

Let,
S: $S = \{s_1, s_2, s_3, ..., s_m\}$ be the sequence of m measurements from sensor X to one sample event;
E: $E = \{e_1, e_2, e_3, ..., e_n\}$ be the sequence of n sample events detectable to sensor X, where each event is described as a k-element set $e = \{v_1, v_2, v_3, ..., v_k\}$ with each element v_i indicating the value of a possible data Quality Influence Factor (QIF) to sensor X.

Quality Profile (QP). of sensor X is denoted as a family of parameterized distributions $P(S|E)$, one for each value in E.

Inverse Quality Profile (IQP). of sensor X is denoted as a family of parameterized distributions $P(E|S)$, one for each possible sensor measurement ever reported to any sample event.

By assuming a Gaussian observation model, $P(S|e_i)$, which is also of more detailed form $P(S|v_1^i, v_2^i, ..., v_k^i)$, can be represented as $N(\mu_i, \sigma_i^2)$, where μ_i and σ_i are the mean and variance of sensor measurements in responding to event e_i; thus, QP of sensor X can be described as a table

$$\{(\mu_1, \sigma_1^2), (\mu_2, \sigma_2^2), ..., (\mu_n, \sigma_n^2)\},$$

one for each sample event. A segment of a sample QP is shown in **Tab.** 3, which describes the behaviour of an ultrasonic motion sensor based on the calibration experimental data shown in **Tab.** 2 with two QIFs *Angle* and *Distance* considered.

By assuming a Gaussian belief model, $P(E|s_j)$, which is also of more detailed form $P(V_1, V_2, ..., V_k|s_j)$, can be represented as $N(\mu_j, \Sigma_j)$, where μ_j is a length-k vector of mean and Σ_j is $k \times k$ matrix of covariances of values of k sensor quality influence factors in responding to a sensor reading s_j; thus, IQP of sensor X can be described as a table

$$\{(\mu_1[k], \Sigma_1[k, k]), (\mu_2[k], \Sigma_2[k, k]), ...\},$$

one for each possible sensor measurement ever reported to any sample event.

2.2   Basic Schemes Using Quality Profile

Sensor behaviour, which is described in sensor QP, can be verified through Goodness-of-Fit (QP_GoF) test or Probability Density Calculation (QP_PDC).

Sensor Behaviour Verification Through Goodness-of-Fit Test. The Chi-square test is used to test if the reported sequence of measurements from a sensor differs from the expected behaviour to the underlying event, which is represented as a parameterized distribution in its QP table, e.g., $N(\mu, \sigma)$. If the Chi-Square sum calculated is less than the probability of exceeding a predetermined critical value, e.g., $\chi^2_{calculated} < \chi^2_{0.01}$, a conclusion can be inferred that the sensor is working with its normal behaviour.

As Chi-square test can only reflect its significance when the testing sample size exceeds certain number, e.g., 20 samples, behaviour verification through Goodness-of-Fit test requires the verifier to buffer a batch of readings from a sensor before it could verify its behaviour. If 20 samples is the minimal requirement for sample size to perform the Chi-square test and there are k sensor QIFs, which is of $(n_1, n_2, ..., n_k)$ number of possible values, in the best case, 20 samples are required to verify the behaviour of a sensor, and in the worst case, $20 \times \prod_{i=1}^{k} n_i$ samples are required to be buffered to perform behaviour verification.

When the space of sensor QIFs is not too huge and the sensors to be verified are generally operate at a rather high sampling rate, the approach of Goodness-of-Fit Test is preferred for sensor behaviour verification as it could provide us more accurate conclusion. In those situations where these two assumptions do not stand, the second approach to perform the verification through Probability Density Calculation is more appropriate to use.

Sensor Behaviour Verification through Probability Density Calculation. When either the buffer size or the buffering time is not affordable at the verifier to use Goodness-of-Fit test, the method to calculate the probability density provides a quick way to verify sensor behaviour. In this approach, the area $A = \int_{s-\epsilon}^{s+\epsilon} f(x)dx$ is used to evaluate per reading based if the sensor works with its normal behaviour; where s is a single reading from the sensor to be verified, $f(x)$ is the probability density function of the distribution corresponding to the underlying event in the QP and ϵ is the predetermined tolerable sensor reading error threshold. As shown in **Fig.** 1, to the same underlying event, grey area (a) represents a better behaviour of a sensor than area (b).

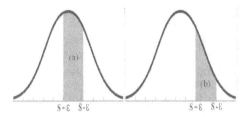

Fig. 1. Sensor Behaviour Verification through Probability Density Calculation

Due to the uncertainty of data in sensor networks, unlike the behaviour verification approach using Goodness-of-Fit test, one run of verification using probability density can only detect out-layer readings, but are not sufficient to indicate abnormal sensor behaviour; and sequential runs of verifications are required.

Underlying Event Estimation. In general situation, the actual underlying event e is unknown to sensors. Therefore, in order to examine behaviour of a sensor, a verifier usually need to make a estimation $\hat{e}$ to the underlying event based on its own measurement s' by using its IQP $P(E|S)$. Such process can be represented as $\hat{e} = \arg\max_{e} P(e|s')$.

The event estimation $\hat{e}$ can also be represented in a more detailed form as $\hat{e} = \{\hat{v_1}, \hat{v_2}, \hat{v_3}, ..., \hat{v_k}\}$ by indicating the value of each possible sensor QIF. It will be chosen from the QP of the sensor to be verified the behaviour under the most similar event to such estimated one for behaviour checking. Given two events e and e', where $e = \{v_1, v_2, v_3, ..., v_k\}$ and $e' = \{v'_1, v'_2, v'_3, ..., v'_k\}$, the similarity between two events is measured by the distance between them as

$$Distance(e, e') = \sum_{i=1}^{k} w_i |v_i - v'_i|$$

where w_i is a predetermined weighting to indicate the significance of a QIF.

2.3   Possible Drawback of Quality Profile

Any fingerprinting technique has a potential drawback that the fingerprint itself as well as the profiling methods are generally public information. Thus, the quality profile of a sensor can be disclosed to the adversary. The adversary can then launch a *imitation attack* using a compromised sensor.

Sensor Behaviour Imitation is a process for a fabricated sensor to estimate the measurement that the original sensor should have generated in responding to the underlying event based totally on the measurement of the fabricated sensor itself. As limited number of sample events are described in quality profile, such process can be represented as

$$P(s|s') = \sum_{i=1}^{M} P(s|e_i)P(e_i|s')$$

where,

s': a particular measurement value from the fabricated sensor,

s: the possible guess on the measurement of original sensor given s',

M: the size of sample event space,

e_i: i^{th} sample event, which is described as a multi-element set based on the number of sensor QIFs,

$P(s|e_i)$: disclosed quality profile of original sensor,

$P(e_i|s')$: inverse quality profile of fabricated sensor

It becomes possible for a fabricated sensor to imitate the behaviour of the original sensor by reporting an estimation $\widehat{s}$ when the actual measurement is s'.

$$\widehat{s} = \arg\max_{s} P(s|e_i)P(e_i|s')$$

In the following example, for simplicity, only one sensor QIF is considered here, e.g., the distance to a sensor, and Gaussian approximation is assumed for both observation model and belief model, which are respectively described in QP and IQP of a sensor. Therefore, the probability density function of ith entry $P(E|s_i)$ in the inverse quality profile of the fabricated sensor will be

$$g_i(x) = \frac{1}{\sigma_{si}\sqrt{2\pi}}e^{-(x-\mu_{sj})^2/2\sigma_{si}^2}$$

And, the probability density function of jth entry $P(S|e_j)$ in the quality profile of original sensor will be

$$f_j(x) = \frac{1}{\sigma_{ej}\sqrt{2\pi}}e^{-(x-\mu_{ej})^2/2\sigma_{ej}^2}$$

Sample events are described as a discrete set $E[M]$ in quality profile, where M is the number of sample events recorded, and for simplicity a constant 2ϵ value step is assumed for the QIF considered. By assuming the sensor readings follows a Gaussian distribution $N(\mu, \sigma^2)$ to an event, 99.7% of readings will fall into the range of $[\mu - 3\sigma, \mu + 3\sigma]$; therefore, in practice, sensor readings can be examined as if a discrete set $S[N]$ with 2θ as the step, where θ is determined by the error tolerable level of sensor readings and N equals to $\frac{3\sigma}{\theta}$. In **Tab. 1**, it is shown a typical sensor behaviour imitation process with one QIF considered is of the complexity O(MN), which is rather low.

Table 1. A Sensor Behaviour Imitation Process

A Sensor Behaviour Imitation Process
//Given the fabricated sensor reading s'
proc Sensor_Behaviour_Imitation(s')
for $i = 1$ to M //
$e = E[i]$
//$g'(x)$ is the pdf of fake sensor's IQP entry to reading s'
$A = \int_{e-\epsilon}^{e+\epsilon} g'(x)dx$
for $j = 1$ to N
$s = S[j]$
//$f_i(x)$ is the pdf of original sensor's QP entry to event $E[i]$
$B = \int_{s-\theta}^{s+\theta} f_i(x)dx$
when $Max(AB) = true$
$\widehat{s} = s$

2.4 Discussions

Since most sensors are not designed to be tamper-proof, our methods proposed above suffer from the *imitation attack*. We hereby discuss the harmfulness of *imitation attack* and point out several techniques to protect the QP.

An adversary can get the QP in two ways: either by taking an existing one or by measuring the QP itself. In the first case, the adversary can not find the QP from the sensor itself (the QPs are normally stored at some trusted places like the base station, not on the sensor.), while it takes much time for the measurement in the second case. In either case, the adversary can not exploit the QP immediately. Noted that when an adversary compromises a sensor with its secret key, he can exploit it immediately. This subtle difference makes it possible for the verifier to detect a small delay when sensors are reporting their readings regularly.

Now we assume the adversary has a disclosed QP of a compromised sensor. He can definitely launch the *imitation attack* for reporting a wrong event with small deviations. There is no effective way of justifying these sensor readings, but by collecting multiple sensor readings (that are not tampered with), the verifier is able to justify the current event by consulting other observers (refer to section 4.2 for the countermeasures). If the adversary only reports the good events (the same events as reported by others) by mimicking the original sensor, we have no effective way to detect it, but accept it as not very harmful. This does happen in certain advanced attack like this: e.g., for reporting sensor readings, a faked sensor launches *imitation attack* to cheat the verifier; for other functions (like forwarding, routing, etc.), it launches other corresponding attacks. Again, we stress that our QP method is only one possible mechanism for defending against limited sources of attacks.

Given disclosed QP, the *imitation attack* needs to buffer some readings and delay a while for computing imitated readings. The defending mechanisms can exploit these facts: the protocol can force sensors to report their readings one by one with no delay, the adversary has no buffer or time to translate the readings. However, sometimes if there is no tight time-synchronization protocol, the method is not very valuable.

3   How to Use QP in a Sensor Network Model

Above we described how to use QP or IQP to verify the sensor readings in a basic scheme. Although the basic scheme can somehow justify the sensor readings, the better if it is used as an assistant mechanism on authenticating the sensor with some other security means. Especially in the case that a large scale of sensors are deployed in a wide area, where they are easily attacked. In this section, we show how the QP methods can be used to protect the sensor readings together with traditional security mechanisms.

Many approaches [8][9][10][11][12] studied secure fusion service in sensor network. We depict a sensor network model as shown in **Fig.** 2, where a cluster head (CH) [1] is responsible for collecting local sensor readings, aggregating and reporting to remote base station. We assume that the sensor network already has the following security mechanisms: proper cryptographic primitives for encryption/decryption and authentication, hierarchical or pairwise key management schemes and secure data fusion protocol. These security mechanisms provide confidentiality, authenticity and integrity for messages sent on the sensor network. Particularly in **Fig.** 2, the cluster head establishes pairwise keys with the base station and each of its local neighbors. All messages between any sending party and receiving party are encrypted and authenticated. To embed our QP methods, the cluster head needs to be equipped with the quality profiles of its neighbors. The base station is responsible for the maintenance of individual quality profiles and updates them periodically to the cluster head.

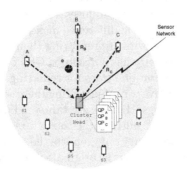

Fig. 2. A network model of using QP. The cluster head is the fusion point covering its local communicating area (the shadow circle), which contains sensors $(A, B, C, S1, S2, ...)$.

Notations:
R_X: A reading of sensor X,
QP_X: The quality profile of sensor X,

---

[1] Typically, a cluster head is selected by possessing higher computation capability and better quality of communication bandwidth.

K_{X-Y}: A pairwise key shared by sensor X and sensor Y,
$ENC\_MAC_K[M]$: Encrypt and authenticate a message M with key K [2],
$DEC\_MAC_K[M]$: Decrypt and authenticate a message M with key K,
The procedure of the secure data fusion protocol is as follows:

1 Suppose the event e is detected by sensor A, B and C. The sensors may generate their reports R_A, R_B, R_C with the following format
$R_i = ENC\_MAC_{K_{CH-i}}[e_i||r_{i1}, r_{i2}, \cdots, r_{ij}]$, where $i \in \{A, B, C, ...\}$ and $j \in \{1, 2, 3, ...\}$.
The reports are then sent to the cluster head.

2 On receiving any report, the cluster head first computes $DEC\_MAC_{K_{CH-i}}$ $[R_i]$ for the decryption and authentication. If successful, the cluster head gets R_i.

3 The cluster head then verifies the data readings $[r_{i1}, r_{i2}, \cdots, r_{ij}]$ according to its sensor quality profile QP_i with either QP_GoF or QP_PDC method.

4 Only verified sensor readings are aggregated together for a cluster head's report R_{CH}.

4   Security Analysis

The basic QP scheme above can determine whether a series of sensor readings come from a sensor. In this section, we analyze various attacks on sensor network, and the possible countermeasures against these attacks. Specially, we may point out the defending ways of using our QP methods.

4.1   Attack Analysis

Sensor networks, like any other network, are suffering from many common attacks, e.g. DDoS attack, replay attack. These attacks have been well studied in the context of Internet, their countermeasures can also be used for sensor networks. For instance, in *resource consumption attack*, an adversary can jam a sensor node by repeatedly sending packets to it, the sensor will soon run out of battery or can not use its radio bandwidth. This is actually a kind of DoS attack [2] on sensor network, which is countered with certain traceback mechanism [5][7]. But as long as the adversary has extra power, unlimited RF bandwidth and computing resources, there is no effective defending mechanism. Another example is the *Replay attack*, where an adversary can replay an old message. Since the message is normally encrypted and authenticated, it should be processed as good message. Thus, the final sensing result may be intentionally modified. Replay attacks, applied on Internet, are countered with the usage of time-stamp and sequence number. These methods can also be used here if there is tight time synchronization or enough sensor memory. We stress that these attacks are mainly defended by traditional security means.

---

[2] The key is split into one encryption key and one authentication key, and used separately.

There are some security approaches [3][4][6] that were proposed particularly for attacks and defenses on different layers of sensor network. The *Wormhole attack and Sinkhole attack*, proposed in [3], are harmful attacks on sensor routing protocols. In *Sybil attack* of [4], an adversary can pretend to be one or multiple legitimate sensor nodes in the sensor network. Several defending techniques for Sybil attacks are: radio resource testing, random key predistribution, registration, position verification and code attestation (refer to [4] for details). However, some of the techniques are still compromised by *Node fabrication attack* (Replacement attack), where an adversary can physically take control of the sensor node. The sensor can be dropped or destroyed, but the worst thing is for the sensor to be used for sending manipulated data. In our work, we don't deal with these attacks directly, but defend against them implicitly (refer to section 4.2).

As mentioned above, we focus on a class of *environmental attacks* (passive or active). A sensor reading can be affected by many factors due to nature condition changes or malicious attacks. Some of them are predictable, e.g. low battery; yet some are unpredictable. Proper operating processes are needed to prevent floating data in case of predictable conditions. While more strict detection mechanisms are needed to prevent active attacks. For those environmental factors, like location, orientation, temperature or noise level, we measure their correct conditions into their quality profiles. Noted that these conditions may be changed passively (due to environmental changes) or actively (due to malicious adversary), we then need to adjust them according to some predefined rules so that the results are generated correctly and promptly.

4.2   Countermeasures Using QP

We set up the game between an attacker and a verifier like this: the attacker sends a series of readings $[r_1, r_2, \cdots, r_j]$ as the input of the verifier and claims that they are coming from sensor "A". The verifier justifies the readings using our QP methods and outputs the final result as $V =< TRUE, FALSE >$, where "TRUE" means the verification is positive, or the readings are coming from the sensor "A", or the attacker wins, and vice versa for "FALSE". Sometimes the verifier needs a known event e [3] as an assistant input. Below, we counter these attacks:

Environmental attack: Given a known event "e", and a series of readings $[r_1, r_2, \cdots, r_j]$ coming from sensor A, the verifier evaluates an event e_A from QP_A. Applying the QP-GoF or QP-PDC method, if $V = TRUE$, We believe that the readings are originated from "A", or to say we lose and the attacker wins the game. If not, we win.

Node fabrication attack: Given a known event "e", and a series of readings $[r'_1, r'_2, \cdots, r'_j]$ from the fabricated sensor A', the verifier generates an event e'_A from QP_A and evaluates it with our QP-GoF or QP-PDC method. If the result is "TRUE", the attacker wins the game, otherwise it loses.

---

[3] A known event can be viewed as a standard test point or other correlated consulting point.

Sybil attack: We study a simple case: a sensor "A" claims its identity at another location called "B" [4]. Given a known event "e" at "B", "A", who is not physically at "B", can either generate arbitrary readings or fabricate sensor "B" and generate "B"'s readings. Suppose we can defend against *environmental attack* and *node fabrication attack* discussed above, we are sure that the attacker can not win in either case.

Wormhole/sinkhole attack: One way of enabling *wormhole attack* is through *Sybil attack*, by claiming a faked identity on the critical route. From above analysis, our QP methods can somehow defend against *Sybil attack*, thus in some implicit way, the methods also help defend against *wormhole attack or sinkhole attack*.

Above countermeasures are only effective if the sensors are profiled accurately for appropriate sensing behaviors. In case the sensors are only involved in the activities such as forwarding messages, routing or aggregating data, or as storage nodes, that are not directly relevant to the data generation phase, our methods may not apply.

5   Conclusion and Future Works

In this paper, we proposed a new approach for improving security of sensor network. We identified a kind of passive fingerprinting technique to be used for profiling sensor behaviours. This profile is then used for verifying the new sensor readings. We defined the concepts of QP and IQP in the context of an ultrasonic motion sensor environment and introduced the basic methods to use them. The preliminary experiments on single sensor are done for concept-proof. Moreover, we demonstrated how to use QP in a network sensor model. The security analysis shows that our methods are useful in defending against various attacks. In the near future, we will acquire correlated sensing models that should be more important in profiling cooperated sensing behaviours.

References

1. Adrian Perrig, John Stankovic, and David Wagner "Security in wireless sensor networks." Communications of the ACM, 47(6), June 2004, Special Issue on Wireless sensor networks, pp.53-57.
2. Anthony D. Wood, John A. Stankovic. Denial of Service in Sensor Networks. IEEE Computer, 35(10):54-62, 2002
3. Chris Karlof and David Wagner, "Secure Routing in Wireless Sensor Networks: Attacks and Countermeasures", First IEEE International Workshop on Sensor Network Protocols and Applications, May 2003
4. James Newsome, Elaine Shi, Dawn Song and Adrian Perrig. "The Sybil Attack in Sensor Networks: Analysis and Defenses." In Third International Symposium on Information Processing in Sensor Networks (IPSN 2004).

---

[4] The location "B" is also regarded as the identity of a sensor "B", by default.

5. Thomas Martin, Michael Hsiao, Dong Ha, Jayan Krishnaswami, Denial-of-Service Attacks on Battery-powered Mobile Computers Second IEEE International Conference on Pervasive Computing and Communications (PerCom'04) March 14-17, 2004 Orlando, Florida pp. 309-318

6. Weichao Wang, Bharat Bhargava. Visualization of Wormholes in Sensor Networks. ACM WiSe 2004, October 1, 2004.

7. Damon Smith, Ryan Mahon, Swathi Koundinya, Shubhashri Panicker. SNTS: Sensor Node Traceback Scheme. ACM WiSe 2004, October 1, 2004.

8. L. Hu and D. Evans. Secure aggregation for wireless networks. In Workshop on Security and Assurance in Ad hoc Networks. Jan. 2003.

9. Wenliang Du, Jing Deng, Yunghsiang S. Han, and Pramod Varshney. A Witness-Based Approach For Data Fusion Assurance In Wireless Sensor Networks. IEEE 2003 Global Communications Conference (GLOBECOM). San Francisco, CA, USA. December 1-5, 2003.

10. B. Przydatek, D. Song, and A. Perrig. SIA: Secure Information Aggregation in Sensor Networks. In Proc. of ACM SenSys 2003.

11. David Wagner Resilient Aggregation in Sensor Networks. ACM Workshop on Security of Ad Hoc and Sensor Networks (SASN '04), October 25, 2004.

12. Joao Girao, Dirk Westhoff, Markus Schneider. CDA: Concealed Data Aggregation in Wireless Sensor Networks. ACM WiSe 2004, October 1, 2004.

Appendix: Sample Quality Profile Segment of an Ultrasonic Motion Sensor

Table 2. Calibration experiments data for an ultrasonic motion sensor with sample events occur at center beam within [2.0, 3.0] meters distance to the sensor

Sample Events v2: distance (m)	Sensor Readings (Angle: $v1 = 0^o$)
2.0	2.0843 2.1968 2.1968 2.2251 2.1785 2.0163 2.0599 2.1562 1.9666 2.0997 2.0846 2.1170
2.1	2.2627 2.2380 2.2380 2.2602 2.3286 2.1787 2.2108 2.1818 2.1005 2.2158 2.2336 2.2525
2.2	2.3170 2.3008 2.3008 2.3157 2.4021 2.2575 2.2792 2.2193 2.1716 2.2783 2.2926 2.3371
2.3	2.3565 2.3533 2.3533 2.3631 2.4745 2.3305 2.4117 2.3225 2.3308 2.3985 2.4219 2.3930
2.4	2.5848 2.4666 2.5278 2.5626 2.5623 2.5459 2.4685 2.4133 2.4169 2.5220 2.5335 2.5409
2.5	2.7223 2.6342 2.7166 2.7105 2.7286 2.7607 2.5409 2.5360 2.5456 2.6982 2.6974 2.5854
2.6	2.7525 2.6787 2.7539 2.8524 2.8565 2.8540 2.7253 2.7097 2.7182 2.7327 2.7317 2.7215
2.7	2.8060 2.7563 2.8041 2.9163 2.9215 2.9086 2.7887 2.7709 2.7813 2.7959 2.7917 2.7558
2.8	2.8420 2.8307 2.8367 3.0272 2.9800 2.9687 2.9166 2.8883 2.8903 2.8392 2.8304 2.8211
2.9	3.1298 3.0897 3.0815 2.8225 2.9890 2.6540 3.0069 2.9893 2.9888 2.9512 2.9012 2.8104
3.0	3.2494 3.2201 3.2124 2.8749 3.1526 2.8749 3.1748 3.1625 3.1581 3.0947 3.0261 2.8617

Table 3. Quality profile segment of an ultrasonic motion sensor with sample events occur at center beam within [2.0, 3.0] meters distance to the sensor

Sample Events		$N(\mu, \sigma^2)$	
v1: angle (^{o})	v2: distance (m)	μ	σ
0	2.0	2.1152	0.0787
0	2.1	2.2251	0.0559
0	2.2	2.2893	0.0578
0	2.3	2.3758	0.0452
0	2.4	2.5121	0.0572
0	2.5	2.6564	0.0831
0	2.6	2.7573	0.0617
0	2.7	2.8164	0.0619
0	2.8	2.8893	0.0696
0	2.9	2.9512	0.1357
0	3.0	3.0885	0.1437

One-Time Sensors: A Novel Concept to Mitigate Node-Capture Attacks

Kemal Bicakci, Chandana Gamage, Bruno Crispo, and Andrew S. Tanenbaum

Department of Computer Science,
Vrije Universiteit Amsterdam, The Netherlands
{kemal, chandag, crispo, ast}@few.vu.nl

Abstract. Dealing with captured nodes is generally accepted as the most difficult challenge to wireless sensor network security. By utilizing the low-cost property of sensor nodes, we introduce the novel concept of one-time sensors to mitigate node-capture attacks. The basic idea is to load each sensor with only one cryptographic token so that the captured node can inject only a single malicious message into the network. In addition, sybil attacks are avoided and explicit revocation is not necessary using one-time sensors. By using public key techniques, one-way hash functions and Merkle's hash tree, we also show efficient implementations and interesting tradeoffs for one-time sensors.

Keywords: Sensor network security, one-time sensor, node-capture attack, sybil attack, Merkle's hash tree.

1 Introduction

Believe it or not, one of the great inventions of humankind was considered as disposable **one-time** baby diapers [1]. Napkins, plastic utensils, cameras are just a few other examples where one-time usage is widespread. The idea of one-time has also found numerous applications in the digital security world e.g., one-time pads, one-time passwords, one-time credentials, etc. While traditionally one-time usage is generally preferred for convenience, regarding security applications most of the time the aim is instead to improve security.

Over the last few years, it has become more clear that the sciences of security and economics have a strong connection between. Within this line of argument we claim that security is a more challenging issue in **wireless sensor networks**, especially because of economic factors. More precisely, since it is required to make the sensor nodes low-cost, they are designed as (1) resource constrained devices and (2) without tamper resistant hardware. The latter deficiency makes sensor nodes which are frequently deployed in unprotected areas vulnerable to "node-capture" attacks while the former limitation puts stringent constraints on the defenses against such attacks.

By using cryptography in the sensors, it is easy to prevent attacks by unauthorized intruders. On the other side, cryptography by itself can not prevent node-capture or inside attacks because in this case the attacker would have the

R. Molva, G. Tsudik, and D. Westhoff (Eds.): ESAS 2005, LNCS 3813, pp. 80–90, 2005.
© Springer-Verlag Berlin Heidelberg 2005

full control over the sensor, including the cryptographic keys on it. Up to now, coping with compromised nodes remains to be one of the most difficult challenges to wireless sensor network security.

In this paper, we present the concept of one-time sensors to mitigate node-capture attacks. The idea is to preload every sensor with a single "cryptographic token" before deployment, so that any node can only insert one legitimate message. Note that for the applications we consider, this message cannot be arbitrary, it rather has a pre-established semantics (e.g., the sensor's reading is above the threshold level or a pre-fixed event has been triggerred).

In our terminology the generic term "cryptographic token" (or just "token") has a different meaning than cryptographic keys and refers to a unique, unforgeable, verifiable and ready-to-transmit bit string. Since there is store-and-forwarding in the wireless sensor network, besides the token, every node is also preloaded with sufficient amount of data to verify others' tokens. This brings us the following advantages:

- The attacker capturing a node can inject at most one malicious message.
- The base station can identify the sources of malicious messages by keeping track of which token was loaded into which sensor.
- Since the tokens are ready to transmit, the sensors do not need to do any cryptographic computation to originate them. However when they are forwarding, the intermediate nodes do require the processing to check the validity of tokens. Note that verification can be implemented more efficiently than generation with some algorithms.
- Previously, explicit revocation was the general strategy to revoke the captured node and prevent energy deprivation attack. One-time sensors eliminate the need for the base station to broadcast a revocation message to the network. We will later show that when one-time sensors are used, the damage an attacker can do with a captured node cannot be more than the cost incurred by sending the revocation message.

Considering these benefits (as well as the low-cost property), we claim that using one-time sensors makes perfect sense for a class of security critical sensor network applications. These include applications where the sensors lose their functionality after the first sensing (e.g., some chemical detectors [2]) or when the sensors can be used only one-time because of external conditions (e.g., fire sensors in the fire scene). When the sensor network should carry alarm messages for rarely-happened events (e.g., nuclear attacks), again one-time sensors are very appropriate.

The rest of this paper is organized as follows. In the next section, we will first give a good application for one-time sensors and then elaborate on its security requirements. In section 3, we will provide a solution based on one-time sensors and discuss three different implementation alternatives. In section 4 we will provide a security analysis. In section 5 we will explore previous studies and their shortcomings. In section 6 we will describe future work and conclude.

2   Application Example and Its Security Requirements

Think of an application where a wireless sensor network is installed in a forest to detect forest fires. In this network possibly thousands of sensors are employed to be able to cover the area concerned. There might be one or more base stations serving as the data sink but for simplicity let us assume there is only one. Unlike sensor nodes, the base station is a powerful and physically protected node.

We assume that the base station knows the locations of sensor nodes at least roughly. Upon detection of a fire in its coverage area, a sensor transmits an alarm message to a base station, which then takes the necessary actions. Our claim is that this example represents an important portion of security critical sensor network applications (e.g., earthquake monitoring, nuclear attack detection, flood detection etc.) in which the common type of communication is between a sensor node and the base station rather than between individual sensors. However since the cheap node has only a small range of transmission, the message it has generated should be forwarded by intermediate nodes (acting as routers) hop by hop until it reaches the base station.

The basic security requirements for this example are as follows:

1. Protection against False Alarms: False alarms should be deterred as they pose a real problem today. Recent investigations [3, 4] found that 98-99% of all security alarms are false alarms. In our case, anyone who would like to insert a false alarm message into the network has 3 choices: (i) as an outside intruder he can generate bogus messages or (ii) replay old intercepted alarm messages (iii) he can capture sensor node(s) and as an insider inject malicious message(s) into the network.
2. Resilience of True Alarms: The network should be designed in a way to assure that true alarms can reach to the base station in spite of attacks.
3. Availability: Since the sensor nodes are battery powered devices, it is important to protect them from denial of service or energy deprivation attacks. Otherwise the sensors receiving, processing and sending unnecessary traffic may lose their functionality very quickly. This might have serious consequences since true alarms cannot be detected by the sensor network any more.
4. Efficiency: Since the sensors are resource constrained devices (i.e. have storage, energy, processing power, transmission constraints), it is crucial to have a solution which has a good performance in terms of all these parameters.
5. Scalability: The solution should work reliably and efficiently when thousands or even tens of thousands of nodes are used.

As you might have noticed, confidentiality is not in our list of security requirements because it is not reasonable to consider the fire information as secret. Also, nonrepudiation is not a requirement.

3   Solution

Traditionally, cryptography is the fundamental tool in the toolbox of security designers. On the other side, the low-cost property of sensors where you do not care much if you throw away some of the nodes is a unique feature of most sensor network applications including the one we discussed here. In our view, to deal with unique security challenges here, it is essential to integrate the counter-measures of **cryptography** and **redundancy** [1] . For this reason we will first discuss briefly how redundancy can help to meet some of the aforementioned requirements and then start explaining our cryptographic solution based on the concept of one-time sensors. We will finish this section by explaining three different implementations of one-time sensors.

3.1   Getting Help from Redundancy

It is easy to protect the network from false alarms produced by outsiders by use of cryptography but cryptography does not help when an attacker captures a node. A recent study showed that all of the information located on sensor's memory can be extracted in less than one minute [5]. With this information, attackers can simply analyze it to capture the cryptographic token and use it to inject a malicious alarm message.

Therefore the proposed defenses against captured nodes should benefit from redundancy. Since the nodes are cheap, it is possible to deploy sensor nodes densely enough to achieve resilience to node capturing up to a threshold value e.g., when the threshold is set to be m, $m - 1$ alarm messages would not be sufficient for the base station to notify an alarm. Therefore the attacker must capture at least m nodes, which requires more work but is not impossible [2].

For the resilience of true alarms, redundancy is crucial. For instance the node sensing the fire should send the alarm using multiple paths so that even when a portion of network is under DoS attack (e.g., using jamming signals), the base station can receive the alarm message.

If energy deprivation attacks take aim to exhaust the battery of only one node (or a few nodes), using enough redundancy we do not need to care much about it. However, the attacks affecting the whole network at the same time should be mitigated by all means.

3.2   Initialization and Operation of One-Time Sensors

We now describe one-time sensors, which can satisfy all the security requirements listed in section 2, if there is enough redundancy in the network.

At the initialization phase, the base station preloads every sensor node with a unique ID value and a single cryptographic token. We will explain different ways for implementing the token in the next subsection. All sensor nodes are also preloaded with sufficient amount of verification data to enable them to check the

---

[1] Note that redundancy is already a widely used technique to achieve other desired properties such as fault tolerance, reliability, accuracy etc.

[2] We will show that one-time sensors prevent the attacker to claim multiple identities (sybil attack [6]). Otherwise by using only a single node, the attacker can succeed to generate m different alarm messages.

validity of tokens received. In every node (including the base station) there is also a memory space reserved to store the revocation list which is initially empty but that will be filled with the ID values of received alarms.

Based on the level of redundancy employed (number of nodes within a certain area), the base station decides on the threshold value for the number of received messages to notify an alarm. It also determines maximum allowable latency (as measured by the time between the receipt of 1st and m^{th} message) before it considers received messages as stale and reinitializes its alarm counter to the value of 0.

Then the operation is performed as follows:

1. Based on its local routing information, the one-time sensor sensing the fire sends an alarm message to the node (or multiple nodes) through which it can reach the base station [3]. The alarm message is basically consists of the ID of the sensing node and its cryptographic token.
2. The node receiving the alarm message first checks by comparing the ID value with the entries in its revocation list whether it has already received a valid alarm message from the same node. If not, it then ensures that the token it received is valid. Only if the token is verified correctly, are two actions taken. First, the alarm message is forwarded to the node(s) on the way to the base station. Second, the ID of the sender node is added to the revocation list for future reference.
3. The second step repeats itself with other nodes until the alarm message is received by the base station. The base station verifies that the alarm message is valid and not already received. Based on the threshold value and the number of previously received fresh messages it either decides to notify an alarm or waits for additional alarm messages to come. The base station does not broadcast a revocation message to the network.

3.3   Implementation of One-Time Sensors

One-time sensors can be implemented either with public key signatures or one-way hash functions. The implementation using one-way hash functions can be improved using Merkle's hash tree. Thus we have three cases in total: (1) Public key based (2) One-way hash based (3) One-way hash based with Merkle's hash tree.

With Public Key Signatures: The base station generates a public key - private key pair. It then signs the unique ID of each sensor with its private key. This signature is in fact the cryptographic token preloaded to each sensor. The public key of the base station is also preloaded to all sensors so that they can verify the signatures (the tokens). When a message is received, its validity can simply be checked by verifying the signature using the base station's public key.

---

[3] We do not discuss how routing is performed in this paper. We refer interested readers to [14].

Considering the last two requirements given in the previous section, public key based implementation makes perfect sense in terms of scalability because of the fact that storage, transmission and computation requirements is constant and does not increase proportionally with the number of nodes in the network. On the other hand, it has problems with respect to efficiency. Since public key operations are expensive, especially for low-cost sensor nodes, this can be a viable alternative only if public key algorithms are implemented very efficiently (e.g. as recently shown in [7]). Note that to reduce the computational requirements on the sensors, in our particular case the sensors use the public key operation only for verification of the signature which can be implemented much more efficiently than signature generation with some algorithms [7].

With One-way Hash Functions: The implementation based on one-way hash functions has two variants. Since the first unoptimized version does not scale well and has problems with respect to storage requirements, we improve it in the second version using Merkle's hash tree. We start explaining with the easier one.

The base station generates a sufficiently long unique random number for each sensor to be deployed. It also precomputes the hash of these random numbers one by one. It preloads each sensor with one of the random numbers as well as hash values of the random numbers loaded to every other node. While they are loaded, the hash values are indexed with the ID values to ease the searching. The base station also keeps track which random number is loaded to which sensor. Now in this case, when a message is received, its validity is checked by computing the hash value of the random number received and comparing it with the hash value stored for the sender's ID value.

In contrast to the previous case, this solution performs very well with respect to computational and transmission requirements but it has a serious storage problem especially when number of nodes is large.

Using Merkle's hash tree: To reduce the storage cost, we can use Merkle's hash tree (MHT) construction [8].

A MHT is a binary tree where a value associated with a node is a one-way hash function of the values of its children. As an example Figure 1 shows the case where the MHT has a height of 4 and the total number of leaf nodes is 16. Each leaf node is identified with an ID value, a positive number incrementing starting from 1 in the left-most node.

For our application this MHT can support up to 16 sensor nodes. Leaf nodes of the MHT contain the hash of the random numbers preloaded to each sensor.

$$h_i = hash(R_i) \tag{1}$$

We use the notation where $h_{i.j}$ denotes the value stored in the parent of all children nodes having the ID values between i and j. As an example, the value stored in the parent of node 1 and 2 can be computed as follows:

$$h_{1.2} = hash(h_1 || h_2) \tag{2}$$

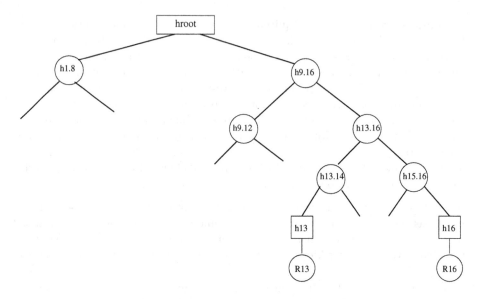

Fig. 1. A Merkle's hash tree supporting 16 sensor nodes

The value of upper parents can be computed similarly. As an example:

$$h_{1.4} = hash(h_{1.2}||h_{3.4}) \tag{3}$$

This continues until we reach the root node.

$$h_{root} = h_{1.16} = hash(h_{1.8}||h_{9.16}) \tag{4}$$

Using MHT, we do not need to load all hash values to all sensor nodes. Instead every sensor node is loaded with (1) a random number (2) the value in the root node (3) the values of the nodes required to recompute the value in the root node.

For instance node 13 is loaded with $R_{13}, h_{14}, h_{15.16}, h_{9.12}, h_{1.8}, h_{root}$ whereas node 16 is loaded with $R_{16}, h_{15}, h_{13.14}, h_{9.12}, h_{1.8}, h_{root}$. These values in fact serve as the cryptographic tokens as well as verification data for these nodes as we will see in an example.

Example: After the deployment, consider the case where node 13 sends an alarm message to node 16 by transmitting its cryptographic token together with its ID value [4]. Node 16 can verify this token as follows:

- It first computes h_{13}.
- It then computes $h_{13.14}$, $h_{13.16}$ and $h_{9.16}$.
- Finally it computes h_{root} and verifies it with the one in its memory.

---

[4] We have chosen node 13 and node 16 to show the worst case. For instance node 16 can verify node 1 using one less hash computation since $h_{1.8}$ is available locally to itself.

Table 1. Comparison of three implementations with respect to network size n

	Storage	Communication	Computation
Public Key Based	$O(1)$	$O(1)$	$O(1)$
Hash Based	$O(n)$	$O(1)$	$O(1)$
Hash Based with Merkle's Tree	$O(\log n)$	$O(\log n)$	$O(\log n)$

As a generalization, we note that for a network having n nodes, using MHT each node has to store only $\log n + 2$ elements instead of n. On the other hand with MHT the communication and computation load increases from the complexity $O(1)$ to $O(\log n)$. Table 1 summarizes the comparison of three implementations described above. There is one other important difference between public key based and hash based solutions. While we can freely add more nodes to the network in public key settings, the maximum size of the network should be determined at the set-up time for the hash based solution.

4   Security Analysis

In this section we briefly analyze the security of one-time sensor implementations.

In both public-key based and one-way hash based implementation, those who are not authorized participants of the network are unable to insert a legitimate alarm message to the network. This is due to unforgeability of the digital signature and one-way property of the hash function used.

Since there is only one cryptographic token inside every sensor, sybil attack [6] is avoided using one-time sensors. In other words, the network is resilient to node-capture attacks up to a threshold value m where m is the minimum number of captured nodes for a successful attack.

Consider the case where sensor nodes are densely deployed and m is chosen sufficiently high. We claim that if the attacker does not know the threshold value m, it is more difficult for him to make the base station to notify a false alarm. Consider the case when the attacker has captured n nodes ($n < m$) and injects n different malicious messages. Just after he realizes that he has not captured enough nodes (for instance by observing that the base station did not take any action), he might want to capture a few more and retry. However since the previous n tokens have already been revoked (assuming that the base station has reinitialized its alarm counter), this does not work. He has to do the capturing again from the scratch and has to try to guess the threshold value in order to decide when to stop.

For our application example, **replay attacks** do not pose a threat for false alarms and the only damage they can cause is energy deprivation. Traditionally protection against replays includes a monotonically increasing counter with every message and rejecting messages with old counter values. Using one-time sensors counters are no longer needed and the revocation table every sensor holds is in fact perfect to protect also against replay attacks. By omitting the counters, only the ID values are needed to be stored therefore storage requirement for the revocation (replay) table is reduced significantly using one-time sensors. However even with this reduction, to fit into sensor node's limited memory we might need to consider further optimizations for the replay table. We do not discuss this issue here any further and leave it as a future work. See [9] for a general discussion of replay protection.

When a sensor transmits its token, its ID is stored in the revocation list of all intermediate sensors forwarding it. Even when the network is designed such a way that multiple redundant paths are used, due to the large size of network, it is reasonable to think that most sensors do not receive the token and store the ID value initially. That is why an attacker who has captured the node or who has intercepted a valid alarm message might replay it in different parts of the network to exhaust the battery of receiving nodes.

One might think that the base station should broadcast a message to the network to explicitly revoke the ID of the node that has sent its token so that the replayed message is not accepted by any node afterwards. However we think this is not so meaningful for two main reasons:

- For security purposes, the revocation messages have to be signed with the base station's private key and every sensor has to verify this public key signature before it updates its revocation table. If one-time sensors are implemented using one-way hash functions, this means sensors should perform public key operations instead of hash computations, which obviously requires much more work and adds to sensor complexity and cost.
- For the case where one-time sensors are implemented with public key signatures, the damage the attacker can do is at most as the cost incurred due to revocation message because in either case, every sensor should receive, process and send a public key signature once. The only exception is bulk revocation where using a single revocation message a number of ID values are revoked together. This latter case is not practical most of the time due to urgent and timely revocation needs of sensor applications for which one-time sensors are proposed.

5   Related Work

In previous work it was repeatedly emphasized that dealing with node-capture attacks is one of the biggest challenge to sensor network security e.g., [10, 11]. According to Perrig et al. [11], "we are a long way from a good solution."

Researchers have recently proposed random key predistribution schemes e.g., [12, 13]. The basic idea is to distribute a random subset of a large pool of symmetric keys to each sensor node. Two nodes that want to communicate first determine whether they share a common secret key, and if they do, then they use the shared key for achieving integrity and confidentiality in their subsequent communication. These schemes are not suitable for our application scenario because it is vulnerable to sybil attacks, meaning that by claiming multiple identities using different keys from the subset the node has, an attacker can even mount a successful attack by capturing only a single sensor node.

For the applications that require end-to-end messaging only between sensor nodes and the base station, one might think that a straightforward solution would be such that each sensor node shares a secret key with the base station. However this solution is vulnerable to energy deprivation attacks because of the fact that the forwarding nodes do not have the capability to verify the legitimacy of messages received.

Using redundancy to protect against node capturing has been previously proposed. For instance in [14], the resilience of multiple paths routing in presence of malicious nodes is analyzed. Additionally, to protect the base station from spoofing, one-way hash functions are used to generate sequence numbers for route discovery messages.

Up to our best knowledge this paper is the first that aims to show how redundancy and cryptography can be used together at a system level to defend against node-capture attacks in sensor networks.

6   Conclusion and Future Work

In this paper we have presented the concept of one-time sensors to mitigate node-capture attacks and three implementation alternatives.

With the novel concept of one-time sensors our attempt is to use the inherent features of sensors and the characteristics of some of their network applications for our own sake of improving the security. One-time sensors are innovative as well as pragmatic if we consider their low-cost property.

Using one-time sensors is not an appropriate choice for applications that require sensors to send arbitrary messages and the integrity and/or confidentiality of these messages should be protected.

As a future work, extending our idea to **k-times sensors** is possible. More formal treatment of redundancy and cryptography integration is highly promising. We are also planning to look at the recent work of improvements on Merkle's hash tree [15, 16] to have a more efficient one-time sensor implementation. Last but not the least, prototype deployments of one-time sensor networks would be very useful to have a better idea on other practical aspects.

References

1. http://www.gpoabs.com.mx/cricher/history.htm
2. CONSESSUS Project, http://www.aramis-research.ch/d/7082.html.
3. http://fox5atlanta.com/iteam/911.html
4. http://www.ci.baltimore.md.us/news/crime/calls.html
5. C. Hartung, J. Balasalle, R. Han: Node Compromise in Sensor Networks: The Need for Secure Systems, Technical Report CU-CS-990-05, Department of Computer Science, University of Colorado, January 2005.
6. J.R. Douceur: The Sybil Attack. In *Proc. 1st International Workshop on Peer-to-Peer Systems (IPTPS'02)*, pages 251-260, LNCS 2429, Springer 2002.
7. G. Gaubatz, J. Kaps, B. Sunar: Public Key Cryptography in Sensor Networks - Revisited. In *Proc. 1st European Workshop on Security in Ad-hoc and Sensor Networks, ESAS 2004*, pages 2-18, LNCS3313, Springer 2005.

8. R. C. Merkle: A Digital Signature Based on a Conventional Encryption Function. In *Proc. Advances in Cryptology - CRYPTO '87*, pages 369-378, LNCS 293, Springer 1988.

9. C. Karlof, N. Sastry, and D. Wagner: TinySec: A Link Layer Security Architecture for Wireless Sensor Networks. In *Proc. 2nd ACM Conference on Embedded Networked Sensor Systems, SenSys 2004*, pages 162-175, November 2004.

10. E. Shi and A. Perrig: Designing Secure Sensor Networks. *IEEE Wireless Communication Magazine*, 11(6), pages 38-43, December 2004.

11. A. Perrig, J. Stankovic and D. Wagner: Security in Wireless Sensor Networks. *Communications of the ACM*, 47(6), pages 53-57, June 2004.

12. L. Eschenauer, V. D. Gligor: A key-management scheme for distributed sensor networks. In *Proc. 9th ACM Conference on Computer and Communications Security*, pages 41-47, ACM 2002.

13. H. Chan, A. Perrig, D. X. Song: Random Key Predistribution Schemes for Sensor Networks. *Proc. IEEE Symposium on Security and Privacy*, pages 197-213, IEEE Computer Society 2003.

14. J. Deng, R. Han, S. Mishra: A Performance Evaluation of Intrusion-Tolerant Routing in Wireless Sensor Networks. In *Proc. IEEE 2nd International Workshop on Information Processing in Sensor Networks*, pages 349-364, 2003.

15. M. Jakobsson, T. Leighton, S. Micali and M. Szydlo: Fractal Merkle Tree Representation and Traveral. In *Proc. Cryptographers' Track at the RSA Conference 2003*, pages 314-326, LNCS 2612, Springer 2003.

16. M. Szydlo: Merkle Tree Traversal in Log Space and Time. In *Proc. EUROCRYPT 2004*, pages 541-554, LNCS 3027, Springer 2004.

Randomized Grid Based Scheme for Wireless Sensor Network

Mohammed Golam Sadi, Jong Sou Park, and Dong Seong Kim

Network Security Lab., Hankuk Aviation University, Korea
{jspark, sadi, dskim}@hau.ac.kr

Abstract. Wireless Sensor Network (WSN) has a wide variety of civil and military applications need enforcement of security. Traditional public key cryptography such as RSA is infeasible due to resource constraints in WSN. Key predistribution is one of the feasible solutions to cope with these constraints. This paper proposes a novel key predistribution scheme named Randomized Grid Based (RGB) scheme which employs the basic probabilistic scheme on the basis of the grid based scheme. Our scheme is not only able to extend resiliency than the existing key predistribution schemes but also ensure a high probability to establish pairwise key and efficiency in path key establishment between sensor nodes. Security analysis shows substantial improvement in term of resiliency and key establishment with little additional overheads in memory, communication.

1 Introduction

Wireless Sensor Network (WSN) consists of a large number of ultra small autonomous devices, called sensor node, powered with battery and equipped with integrated sensors, data processing capabilities and short range radio communication. In typical application scenarios sensor nodes are spread randomly over the terrain under scrutiny and collect sensor data. Resource constraints in the sensor node have established popular perception that traditional public key cryptography and key distribution center are beyond the capabilities of sensor nodes. Although the software implementation of ECC for 8 bit CPUs showed the possibility to take advantage of public key cryptography to constrained devices such as embedded system [5], it is still not feasible for WSN. In order to ensure security to WSN, key predistribution schemes have been widely accepted since the early stage of WSN development. Several studies [1-4, 6] have proposed key predistribution schemes. Eschenauer and Gligor have proposed the basic probabilistic scheme [4]. The main idea is to let each sensor node randomly pick a set of keys from a key pool before deployment. Any two nodes have a certain probability to share at least one key that act as the secret key between them. Chan *et. al.* further extended this idea and developed two key predistribution schemes [2]: q-composite key scheme and random pairwise keys scheme. The q-composite key scheme also uses key pool but requires two sensors compute a pairwise key from at least q-predistributed shared keys. The random pairwise keys scheme picks pair of sensors and assigns each pair a unique random key. Both these

R. Molva, G. Tsudik, and D. Westhoff (Eds.): ESAS 2005, LNCS 3813, pp. 91–101, 2005.
© Springer-Verlag Berlin Heidelberg 2005

schemes improve the resiliency over basic probabilistic scheme. However the basic probabilistic and q-composite key scheme provides very poor performance when the number of compromised nodes increases. The random pairwise keys scheme overcomes the above problem but it needs much memory requirement. Liu and Ning have proposed two efficient schemes [6]: random subset assignment and grid based key predistribution scheme that have basis of polynomial key predistribution. In grid based scheme a conceptual grid is formed and a unique polynomial function is allocated to each row and column of the grid. A sensor node is allocated to a particular intersection of the grid and the two polynomial shares corresponding to that row and column are assigned to its memory. If any two sensors have same column or row number then certainly they can establish a pairwise key if they are in a communication range. This scheme has a number of nice facilities such as high probability to establish pairwise keys, resiliency to node capture, low communication overhead and reduced computation in the sensor node. However, the resiliency to node capture is not acceptable when the compromise of nodes grows larger than certain threshold value.

In this paper, we propose a novel key predistribution scheme named Randomized Grid Based (RGB) scheme to solve the problem of resiliency against the large number of node capture. We employ the basic probabilistic scheme on the basis of the grid based scheme. The combined effect of these two basic schemes improves the network resiliency to a higher level against large number of node capture. Besides this improvement, RGB scheme guarantees a very high probability to establish pairwise keys between neighboring nodes in the absence of compromised nodes. Although some nodes in the network are compromised, there are are several ways to reestablish pairwise keys between the noncompromised nodes through path discovery method.

2   Overview of Randomized Grid Based (RGB) Scheme

In this section we present our proposed scheme in detail. We use grid based scheme as the building block to achieve enhanced resiliency against large number of node capture. Motivated by the probabilistic random key predistribution scheme, we combine probabilistic scheme with the grid based scheme. We construct a $m \times m$ grid structure where $m = \sqrt{N}$ and N is number of nodes in the network. Practically the value of N is chosen larger than the actual number of sensor nodes in the network to keep the option to increase the network size in future if required. We allocate distinct multiple polynomials to each row and column of the grid. An intersection in the grid has been selected for each sensor node and then τ polynomials are randomly chosen from the corresponding row and column of the intersection for that node. Finally these 2τ polynomial shares and the coordinate of the intersection as the sensor ID are assigned to that sensor node. We encode the coordinate of a sensor into a single valued ID. Let $l = \lceil log_2 m \rceil$. Any valid column or row coordinate can be represented as an l bit binary string. We then denote the ID of a sensor as the concatenation of the binary representations of the row and column coordinate values. Syntactically, we represent an ID constructed from the coordinate (i, j) as $\langle i, j \rangle$. For ease of presentation, we denote ID i as $\langle r_i, c_i \rangle$, where r_i and c_i is the first and last l bits of i respectively. If two sensor nodes share a common polynomial, then they can establish a pairwise key between them. The details of the RGB scheme are presented below.

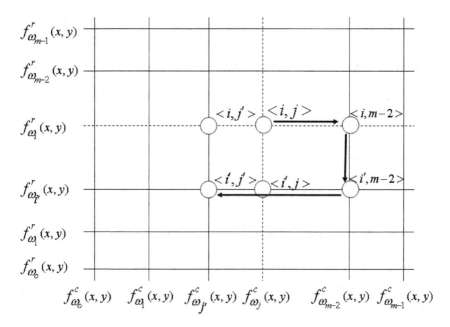

Fig. 1. Polynomial allocation and key discovery mechanism of RGB scheme

2.1   Pre-assignment of Polynomials

Before deployment of the sensors in the practical field the setup server does the following works:

- Randomly generates $2m\omega$ number of t degree bi-variate polynomials $F = \{f_i^r(x,y), f_i^c(x,y)\}_{i=0,\ldots,m\omega-1}$ over a finite field F_q.
- Divides the polynomials $f_i^r(x,y)_{i=0,\ldots,m\omega-1}$ into m groups $f_{\omega_i}^r(x,y)_{i=0,\ldots,m-1}$ where each group contains ω distinct polynomials and assigns to each row of the grid.
- In a similar way the setup server divides the polynomials $f_i^c(x,y)_{i=0,\ldots,m\omega-1}$ into m distinct groups $f_{\omega_i}^c(x,y)_{i=0,\ldots,m-1}$ that are allocated to each column of the grid as shown in Fig. 1.
- For each sensor, the setup server picks an unoccupied intersection (i, j) in the grid and selects τ polynomials randomly from each of the allocated polynomial group corresponding to i^{th} row and j^{th} column of the intersection.
- Finally assigns these 2τ polynomial shares with their IDs and the ID $\langle i, j \rangle$ to the sensor node.

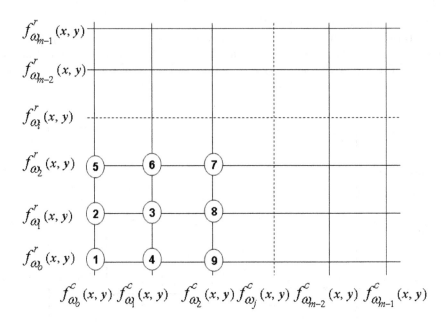

Fig. 2. An example of order of node assignment in RGB scheme

To facilitate path discovery, we require that the intersection allocated to each sensor be in densed rectangle area in the grid. Fig. 2 shows a possible order to allocate intersections to the sensors.

2.2 Discovery of Polynomial Shares

To establish a pairwise key with node j, node i checks whether $c_i = c_j$ or $r_i = r_j$. If $c_i = c_j$, then it is confirmed that both nodes i and j share τ polynomials from the polynomial group $f^c_{\omega_i}(x, y)$. So they go through to discover a common polynomial by simply broadcasting the polynomial IDs in clear text or by using challenge response protocol similar to basic probabilistic scheme [4]. If any common polynomial ID is found then they can establish a pairwise key directly using that common polynomial. Similarly, if $r_i = r_j$, they have polynomial shares from $f^r_{\omega_i}(x, y)$ and go through to establish a pairwise key in similar way stated earlier. If they fail then they have to establish a pairwise key using path key discovery method.

2.3 Discovery of Path Key

Nodes i and j have to use path discovery when $c_i \neq c_j$ and $r_i \neq r_j$ or when there is no common polynomial in their memory. We note that either node $\langle r_i, c_j \rangle$ or $\langle r_j, c_i \rangle$ can establish a pairwise key with both the nodes i and j as they have τ polynomials selected randomly from a common polynomial group. Indeed, if there is no compromised node, it is obvious that there might exist at least one node that can be used as an intermediate node between any two sensor nodes due to the node assignment algorithm. For example, in Fig.1, both nodes $\left\langle i^{'},j\right\rangle$ and $\left\langle i,j^{'}\right\rangle$ can help node $\langle i,j \rangle$ to establish a pairwise key with node $\left\langle i^{'},j^{'}\right\rangle$. Note that nodes i and j can predetermine the possible intermediate nodes without communicating with others.

In some situations, it may be possible that both of the above intermediate nodes have been compromised or are out of communication range. However, there are still alternative key paths. For example, in Fig. 1, beside node $\left\langle i^{'},j\right\rangle$ and $\left\langle i,j^{'}\right\rangle$, node $\langle i,m-2 \rangle$ and $\left\langle i^{'},m-2\right\rangle$ can work together to help node $\langle i,j \rangle$ to setup a common key with $\left\langle i^{'},j^{'}\right\rangle$. Thus all the nodes that belong to same row or column of the nodes i and j can help to setup a pairwise key between them. Indeed, there are up to $2(m-2)$ pairs of such nodes in the grid.

In general, we can map the set of noncompromised nodes into a graph, where each vertex in the graph is one of the sensors, and there is an edge between two nodes if these two sensors have polynomial shares of a common polynomial. Discovering a key path between two nodes is equivalent to finding a path in this graph. Nevertheless, in a large sensor network, it is usually not feasible for a sensor to store such a graph and run a path discovery algorithm. Thus, in our scheme, we focus on the key paths that involve two intermediate nodes. Specifically, a sensor node S may use the algorithm in [6] to discover key paths to sensor D that have two intermediate nodes.

3   Security Analysis of RGB Scheme

We assume that there are $N = m \times m$ sensors in the network. According to polynomial distribution in WSN, each sensor node has a very high probability to establish a pairwise key with $2(m-1)$ sensor nodes directly. Thus, among all the other sensors, the percentage of nodes that a node can establish a pairwise key directly is,

$$\frac{2(m-1)}{N-1} \approx \frac{2(m-1)}{m^2-1} = \frac{2}{m+1} \tag{1}$$

This scheme has reasonable memory requirements mainly for storing several polynomials chosen randomly. Each sensor needs to store 2τ polynomials of degree t and also their corresponding IDs. Assume b bits are required to represent a polynomial ID. Then we can write $b = log_2\,(2m\omega)$. In addition, a sensor needs to store the IDs of the compromised nodes with which it can establish a pairwise key directly. Thus, the total storage overhead in each sensor is at most

$$Memory = 2\tau\,(t+1)\,logq + 2\tau b + 2\,(t+1)\,l \tag{2}$$

First we will compute the desired values for security parameter and then focus our attention to the performance of the RGB scheme under two types of attacks. First, the attacker may target the pairwise key between two particular sensors. The attacker may either try to compromise the pairwise key or prevent the two sensor nodes from establishing a pairwise key. Second, the attacker may target the entire network to lower the probability that two sensors may establish a pairwise key or to increase the cost to establish pairwise keys.

3.1   Preferred Values of ω, τ

For any pair of nodes, establishing a pairwise key between them is possible if the key sharing graph of the nodes is connected. Given the size and the density of a network, we can calculate the values for ω and τ so that the node graph is connected with high probability. We use the following approach adapted from [4].

Computing Required Local Connectivity. Let p_c be the probability that the key sharing graph is connected. We call it global connectivity. We use local connectivity to refer to the probability of two neighboring nodes sharing at least one polynomial. To achieve a desired global connectivity p_c, the local connectivity must be higher than a certain value; we call this value the required local connectivity, denoted by $p_{required}$. Using connectivity theory in a random graph by Erdos and Renyi [6], we can obtain the necessary expected node degree d (i.e. the average number of edges connected to each node) for a network of size N, when N is large in order to achieve a given global connectivity, p_c:

$$d = \frac{N-1}{N} \left[ln(N) - ln(-ln(p_c)) \right] \tag{3}$$

For a given density of sensor network deployment, let n be the expected number of neighbors within wireless communication range of a node. Since the expected node degree must be at least d as calculated above, the required local connectivity $p_{required}$ can be estimated as:

$$p_{required} = \frac{d}{n} \tag{4}$$

Computing Actual Local Connectivity. After we have selected values for ω and τ, the actual local connectivity is determined by these values. We use p_{actual} to represent the actual local connectivity, namely p_{actual} is the actual probability of any two neighboring nodes sharing at least one polynomial share (i.e. they can find a common key between them). Since $p_{actual} = 1 - p_r$ (two nodes do not share any polynomial),

$$p_{actual} = 1 - \frac{\binom{\omega}{\tau}\binom{\omega-\tau}{\tau}}{\binom{\omega}{\tau}^2} = 1 - \frac{((\omega-\tau)!)^2}{(\omega-2\tau)!\omega!} \tag{5}$$

The values p_{actual} have been plotted in Fig. 3 when ω varies from τ to 100 and $\tau = 2, 4, 6, 8$. For example, when $\tau = 6$, the largest ω we can choose while achieving the local connectivity $p_{actual} \geq 0.7$ is 35.

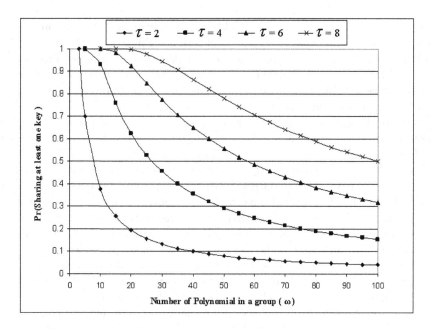

Fig. 3. Probability of sharing at least one key between nodes

Computing ω and τ. Getting the required local connectivity $p_{required}$ and the actual local connectivity p_{actual}, in order to achieve the desired global connectivity p_c, we should have $p_{actaul} \geq p_{required}$,

$$1 - \frac{((\omega - \tau)!)^2}{(\omega - 2\tau)!\omega!} \geq \frac{N-1}{nN}[ln(N) - ln(p_c)] \qquad (6)$$

Therefore, in order to achieve certain p_c for a network of size N and the expected number of neighbors for each node being n, we just need to find values of ω and τ, such that Inequality (6) is satisfied.

3.2   Attacks Against a Pair of Sensors

To explain the security of our scheme, we like to focus the difficulty to compromise a pairwise key without compromising the related nodes and the difficulty to prevent two nodes from establishing a pairwise key. Assume two nodes u and v can establish a pairwise key directly. The only way to compromise the pairwise key without compromising the related nodes is to compromise the shared polynomial between them. It requires the attacker to compromise at least $t+1$ sensor nodes having same polynomial shares and belongs to the same row or column of the target node. Even if the attacker successfully compromises the polynomial (as well as the pairwise key), the related sensors can still reestablish another pairwise key through path discovery process or by using one of the noncompromised polynomials stored in their memory. From the path discovery process, we know that there are $m-1$ pair of nodes, which can help nodes u and v to reestablish a pairwise key. To prevent node u from establishing a key with node v completely attacker has to compromise all of the pair of nodes otherwise there will be a possibility to establish a pairwise key between them through multiple rounds of path discovery process. Thus, in this case, the attacker has to compromise $t + 1$ nodes from the same row or column to learn the preestablished pairwise key and $t + 2m$ sensors to prevent u and v from establishing another pairwise key.

Now we will discuss the scenario where nodes u and v establish a pairwise key through path key establishment. The attacker may compromise one of the sensors involved in the key path. If the attacker has the message used to deliver the key, he/she can recover the pairwise key. However, the related sensors can establish a new key with a new round of path key establishment once the compromise is detected. To prevent the sensors from establishing another pairwise key, the attacker has to block at least one sensor in each path between u and v. There are $2m - 2$ key paths between u and v that involve one or two intermediate nodes. Besides the key path with the compromised node, there are at least $2m - 3$ paths. To prevent pairwise key establishment, the attacker has to compromise at least one sensor in each path. Still there is a probability of reestablishment of new path key having more than two intermediate nodes. Thus, in summary, the attacker has to compromise one sensor involved in the path key establishment to compromise the pairwise key and at least $2m - 3$ sensors to prevent u and v from establishing a pairwise key.

3.3 Attacks Against the Network

Having the knowledge of the subset assignment mechanism, adversary may compromise the bivariate polynomials in F one after another by compromising selected sensor nodes in order to finally compromise the whole network. Suppose the adversary just compromised l bivariate polynomials in F. In the grid based scheme [6] there are about ml sensor nodes where at least one of their polynomial shares has been disclosed. But in our scheme, each node in the same column or row has random selection of polynomials. There is a probability that these compromised polynomials may belong to each sensors of the same column or row and depends on ω and τ. So the polynomials belong to the sensor nodes are disclosed rather than the all the nodes in the same column or row and still those nodes can work normally as they have other uncompromised polynomials in their memory. However we see that adoption of randomness in the bloom's scheme [4] enhances the resiliency to node capture. Now we will calculate the probability of at least one key is disclosed when N_c nodes are captured. An adversary may randomly compromise sensor nodes to attack the path discovery process. It follows from the security analysis in [1] that an attacker cannot determine noncompromised keys if he or she has compromised more than t sensor nodes. We assume that an attacker randomly compromises N_c sensor nodes, where $N_c > t$. Consider any polynomial f in F. The probability of f being chosen for a sensor node is $\frac{\tau}{m\omega}$ and the probability of this polynomial being chosen exactly k times among N_c compromised sensor nodes is,

$$P(k) = \frac{N_c!}{(N_c - k)!k!}(\frac{\tau}{m\omega})^k(1 - \frac{\tau}{m\omega})^{N_c - k} \tag{7}$$

Thus, the probability of any polynomial being compromised is $P_c = 1 - \sum_{t=0}^{k} P(k)$. Since f is a polynomial in F, the fraction of compromised links between noncompromised sensors can be established as P_c. Fig. 4 shows the relationship between the fraction of compromised links for noncompromised sensors and the number of compromised nodes for a combination of $\omega = 25$ and $\tau = 6$. According to the graph, we see that our scheme has a very high security performance when a large number of the sensor nodes are compromised. For example, in the case of a sensor network of 20,000 nodes, if the attacker compromises 40% of the total nodes (i.e. 8,000 nodes) then only about 5% of the links of noncompromised nodes are affected. Thus, the majority of the noncompromised nodes are not affected.

3.4   Comparison with Previous Schemes

Let us compare the RGB scheme with basic probabilistic scheme, q-composite scheme and grid based scheme. Here we assume the network size N is 20000, $m = 142$ and the probability $p = 0.24$. In Fig. 4, the four curves show the fraction of compromised link as a function of the number of compromised sensor nodes. Basic probabilistic scheme has almost same performance as the q-composite scheme ($q = 2$) and the grid based scheme works well up to 2000 compromised nodes. In contrast, our scheme provides sufficient security up to 9000 compromised nodes and then the performance gradually decreases. Here we assume the value of the security parameters $\omega = 25, \tau = 6$ and degree of polynomial $t = 19$.

Now we compare the memory requirements of our scheme with grid based scheme. According to grid based scheme [6], the storage overhead in each sensor is at most $(t + 1)logq + 2(t + 1)l$. Memory overhead of our scheme indicated by Equation (2) mainly depends on the value of the security factor τ and the degree of polynomial t. A comparison of these two equations indicates that our scheme requires almost τ times more memory than the grid based scheme. Fig. 3 describes how the probability of sharing at least one key among nodes varies for different values of τ. Larger value of τ will increase the memory overhead in our scheme. But we can regulate the value of τ considering the network security requirement as stated in Sec. 3.1. To get the performance shown in Fig. 4 each sensor needs the storage capacity, which is equivalent to store almost 260 keys. Though it needs more memory, it is not a significant factor for WSN because size of memory on a sensor node will be increased in near future as technology is developing very fast. In terms of communication overhead, our scheme has additional overhead compared to the grid based scheme due to broadcasting of polynomial IDs during direct key establishment process. But this additional communication will occur only once during the key discovery process and after that stage the communication overhead is similar to the grid based scheme. So the overall communication is not a significant overhead for the sensors. The computational overhead is essentially the evaluation of one or multiple t-degree polynomials that follow the same approach as in [6]. From the above comparison, we can explicitly say that our scheme has a substan-

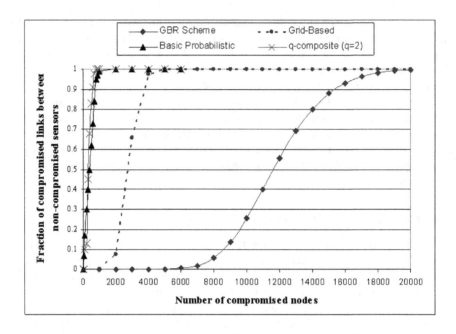

Fig. 4. Fraction of compromised links between noncompromised sensors v.s. number of compromised sensor nodes

tial improvement in network resiliency with little increase in memory usage and reasonable communication and computation workload. It also includes other facilities. First, it guarantees the establishment of pairwise key between two nodes directly or via intermediate nodes when there are no compromised nodes and also provides efficiency in determining the path key. Secondly, if some nodes are compromised there is still a high probability to establish a pairwise key between noncompromised nodes. Finally, this scheme allows optimized deployment of sensors to establish pairwise key directly due to orderly assignment of grid intersections.

4   Conclusions

We have presented a new pairwise key distribution scheme named Randomized Grid Based (RGB) scheme for WSN. Security analysis demonstrates substantial improvement in resiliency against large number of node capture and significant enhancement for establishing a pairwise key between nodes in an efficient way. Furthermore, our scheme enables one to adjust the probability of key establishment by regulating the security parameters according to the desired level of security for WSN. The future works include further analysis on computation workloads in terms of energy requirements through both analytical method and simulation in detail.

Acknowledgement

This research was supported by the Internet information Retrieval Research Center (IRC) in Hankuk Aviation University. IRC is a Regional Research Center of Gyeonggi Province, designated by ITEP and Ministry of Commerce, Industry and Energy.

References

1. Chan, H., Perrig, A., Song, D.X.: Random Key Predistribution Schemes for Sensor Networks. In Proc. of the 2003 IEEE Sym. on Security and Privacy (2003) 197
2. Du, W., Deng, Jin., Han, S.Y., Varshney, P.K.: A Pairwise Key Pre-distribution Scheme for Wireless Sensor Networks. In Proc. of the 10th ACM conf. on Computer and Communications Security. (2003) 42-51
3. Eschenauer, L., Gligor, V.D.: A Key-Management Scheme for Distributed Sensor Networks. In Proc. of the 9th ACM Conf. on Computer and Communications Security. (2002) 41-47
4. Gupta, V., Millard, M., Fung, S., Zhu, Yu., Gura, N., Eberle, H., Shantz, S.C.: Sizzle: A Standards-Based End-to-End Security Architecture for the Embedded Internet. In Proc. of 3th IEEE Int. Conf. on Pervasive Computing and Communications. (2005) 247-256
5. Liu, D., Ning, P.,:Establishing Pairwise Keys in Distributed Sensor Networks. In Proc. of the 10th ACM Conf. on Computer and Communications Security. (2003) 52-61
6. Erods, P., Renyi, A.: On Random Graph. Publicationes Mathematicae. (1959) 290-297

Influence of Falsified Position Data on Geographic Ad-Hoc Routing

Tim Leinmüller[1], Elmar Schoch[1], Frank Kargl[2], and Christian Maihöfer[1]

[1] DaimlerChrysler AG, Research Vehicle IT and Services
{Tim.Leinmueller, Elmar.Schoch, Christian.Maihoefer}@DaimlerChrysler.com
[2] University of Ulm, Department of Media Informatics
Frank.Kargl@informatik.uni-ulm.de

Abstract. There has been a lot of effort in the research on routing in mobile ad hoc networks in the last years. Promising applications of MANETs, e.g. in the automotive domain, are the drive for the design of inter-vehicle networks. So far, several projects in this field have chosen geographic routing approaches because of their outstanding performance and the possibility to support location-based applications like traffic warning functions. Having reached a reasonable functional level, a next step will be a deeper study of safety and security issues.

With this paper, we dive into that area by assuming defective or malicious nodes that disseminate wrong position data. First, we have a look at the local problems that may arise from falsified position data, then we show the global effects on the routing performance by simulating malicious nodes. Simulation results show that the overall ratio of successfully delivered messages decreases, depending on the number of maliciously acting nodes, even up to approximately 30%. We conclude from this result that future work should take these threats into account in order to design more robust routing protocols.

1 Introduction

In the recent years, Mobile Ad hoc Networks (MANETs) have attracted a lot of attention in the research community. Still, there are very few real application scenarios where the wide deployment of MANETs is really foreseeable in the near future. Two exceptions are the military area and networks that spontaneously connect vehicles on the road, so called Vehicular Ad hoc Networks (VANETs). In the latter case, a number of research projects produced significant results concerning routing and other operational issues [1]. Main target of these projects is the improvement of vehicle safety by means of inter-vehicle communication. So e.g. in the case of an accident, a VANET might be used to warn approaching cars and give the drivers enough time to come to a halt. Another application area is using VANETs for entertainment purposes, allowing e.g. news exchange between passengers of different cars.

---

[1] e.g. projects like Fleetnet [1] or CarTalk2000 [2].

R. Molva, G. Tsudik, and D. Westhoff (Eds.): ESAS 2005, LNCS 3813, pp. 102–112, 2005.
© Springer-Verlag Berlin Heidelberg 2005

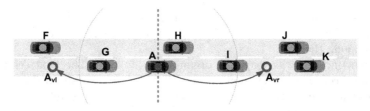

Fig. 1. Vehicle A pretends to be at positions A_{vl} and A_{vr} and thus manages to grab all traffic along the road.

Now European and US car manufacturers are taking the next step in projects that aim at defining a reference architecture and suitable standards for VANETs[2].

In contrast to generic MANETs, where mostly topology-based routing protocols are being developed, many of the VANET projects use position-based routing mechanisms [6] for establishing connectivity between vehicles. This offers some advantages in performance and the possibility to address vehicles by their position (so called Geocast) instead of their address.

Whereas a lot of effort was already put in securing traditional MANETs [7,8], the security research for position-based routing and VANETs is still in its infancy. [9] gives a first overview on this subject. When using position-based routing, one important aspect is the correctness of position data. The routing mechanisms proposed so far all work the same: nodes measure their location by means of some sensors (e.g. GPS) and then distribute the measured location to other nodes which can then base their routing decision on the location of others.

When false position information is distributed in the VANET, this can severely impact the performance of the network, as we will show in this paper. A potential source for such false position data is a malfunction of a node's location sensing system. E.g. a GPS receiver may wrongly calculate the position of a node because of bad reception conditions.

Whereas malfunctioning nodes may degrade the performance of a system to some extent, malicious nodes may cause even more harm. The intents of an adversary may range from simply disturbing the proper operation of the system to intercepting traffic exchanged by ordinary users, followed by a potential modification and retransmission. If the data is not protected, e.g. by cryptographic means, this can lead to a compromise of nearly all security goals like confidentiality, authenticity, integrity, or accountability.

Figure 1 shows a scenario where node A claims to be at two additional (faked) positions A_{vl} and A_{vr}. Based on a greedy forwarding strategy, nodes always select the node nearest to the destination as the next forwarding node. Assuming that F wants to send a packet to node K, it will first sent the packet to its only direct neighbor G. G will then forward the packet to the node nearest to the destination from which it can hear beacons. This seems to be A_{vr}, so the packet ends up at node A, which can now forward, modify or discard it at will. In the

---

[2] e.g. the US Vehicle Safety Communication Consortium (VSCC) [3], the Network on Wheels project (NoW) [4], or the Global Systems For Telematics (GST) [5] initiative.

opposite direction, the packet from K will go to I, which will again send it to the assumed best node A_{vl}. So faking only two positions, A is able to intercept all traffic along the road.

The remainder of this paper is organized as follows. The next section will give a more complete discussion on the effects of false position data. Section 3 provides our simulation results. In section 4 we discuss related work and section 5 concludes our work.

2   Effects of False Position Data

If we assume that false position data is generated by malfunctioning or malicious nodes, what are the possible effects?

Figure 2 shows some of the effects that can occur. If a node's real position is not in the route from source to destination and neither is the false position, then no effect occurs (6). The same is true if real and false position are in the route, but the positions are similar and the position within the route does not change (9).

A node that does not want to be used for forwarding, e.g. to save own resources like energy, bandwidth, etc, may choose to fake a position outside the route (4). Depending on whether there is a backup path (7) or not (8), either packets get lost or at least the routing becomes non-optimal.

Finally, the cases below position (10) can either be reached, if real and false position are both in the route but at different positions, or if the real position is not in the route but the false is. Then one has to distinguish, if the node can receive the packet sent to the false position at his real position (12) or not (11).

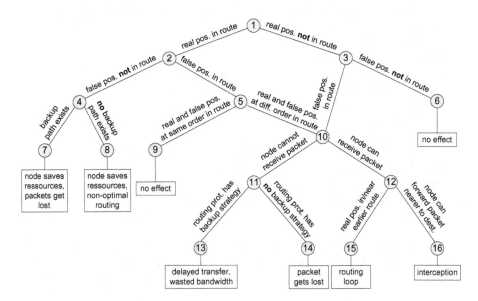

Fig. 2. Possible effects of false position data

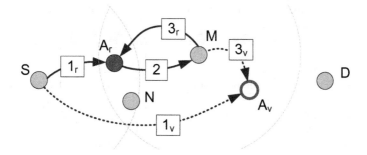

Fig. 3. Routing loop induced by the malicious node A which pretends to be at A_v rather than at its actual position A_r

In case (11), the packet sent to the node is lost. If the routing protocol notices that (e.g. by means of acknowledgments or timeouts) and has a backup strategy (13), the packet may still be delivered to the destination. This will create an additional delay and waste bandwidth, as the first transmission gets lost. If the routing protocol has no such backup strategy, the packets get lost (14).

In case (12), the node receives the packet. If the real position of a node allows the packet to be delivered to a position which is nearer to the destination than the false position (16), then the packet will reach its destination. The benefit of an attacker might be, that he can intercept traffic that would otherwise be routed around him, sniffing e.g. confidential information or similar.

If the real position of the node is further away from the destination than the false position (15) and the node will then forward the packet so it reaches the false position again, routing loops can occur as shown in figure 3. Here node A claims to be at position A_v where its real position is A_r. S sends the packet to the node in its neighborship that claims to be nearest to D (1_v). In reality, node A receives the packet (1_r). It then forwards the packet to node M (2) which again tries to forward it to the node that is nearest to D (3_v). This is the virtual position A_v and so the packet is again received by A (3_r). The steps 2 and 3_r repeat forever or until a time-to-live counter expires.

As we have shown, false position data is clearly an issue that can affect the performance, reliability and security of a MANET using position-based routing. In the next section we will use simulations to show, how severe this impact can be for certain scenarios.

3 Simulative Analysis

3.1 Simulation Environment

In order to be able to estimate the impact of falsified position data on geographic routing, we have implemented position faking in the ns-2 simulation environment. For the routing scheme, we choose a greedy based approach. It selects the neighbor node as next hop for a packet, whose distance to the destination is minimal. Like all greedy methods, this algorithm fails if no neighbor is

Table 1. Short overview on simulation parameters

Parameter	Value
Number of nodes	100
Length of square node field	1000 – 4000m
⇒ node density (nodes/km^2)	6,25 – 100
Max. node velocity (m/s)	50
Pause times (s)	0.0
Mobility model	Random Waypoint
Link-/MAC-Layer	IEEE 802.11
Transmission range (m)	250
Number of sent messages	100
Simulation time (s)	40
Simulation runs	20

found that is closer to the destination than the current node itself. The deployed recovery strategy is based on a caching approach, i.e. packets are stored locally until either a suitable neighbor is reachable or until the node decides to drop the packet (see [10]).

Besides ordinary routing, we also have to integrate a model of maliciously acting nodes. Therefore, a certain percentage of all nodes in the simulation scenario behaves as follows:

1. Whenever a malicious node is about to send a beacon message to announce its present position, it selects a random position on the field and applies it to the beacon (instead of its real position).
2. Whenever a malicious node gets a data packet, depending on the simulation setup, it either forwards it correctly according to the protocol rules or it drops the packet.

As data traffic, 100 messages are transmitted from a random source to a random destination. The messages are randomly created during the first 30 seconds of the simulation run. Further simulation parameters are listed in table 1. Node density, velocities and mobility model approximately reflect the movement patterns of vehicular traffic in an urban area [11].

The following subsections present and discuss our simulation results regarding the impact on ad hoc network routing performance. We take a look at the impact on the delivery ratio and the reasons for the impact, namely parameters such as number of packet drops due to routing loops and number of packets remaining in the routing caches.

3.2   Impact on Delivery Ratio

The influence of falsified position information on the overall number of successfully delivered messages has been measured in several simulation runs with different percentages of position faking nodes. Figures 4 and 5 contain the results of simulation runs in a $2000m * 2000m$ sized network field with 10% and 40% faking

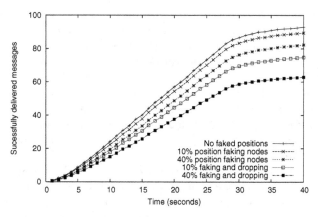

Fig. 4. Successfully delivered messages accumulated over simulation time

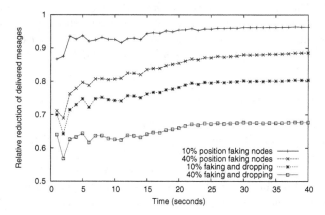

Fig. 5. Relative reduction of successfully delivered messages over simulation time

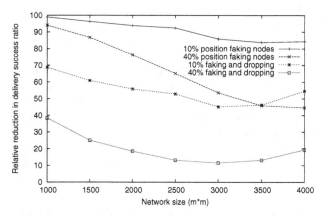

Fig. 6. Relative reduction of successfully delivered messages in dependence of network size

nodes, once with and once without packet dropping. In figure 4, the percentage of successfully delivered messages in total is depicted, whereas figure 5 shows the relative decrease compared to the case without falsified position information.

As expected, with position faking, the delivery ratio is always negatively influenced. In case faking nodes do also drop received packets, the impact is even more severe (see figure 4). The relative comparison in figure 5 shows, after an initial phase, pure position faking decreases the overall delivery ratio by approximately 4% for 10% faking nodes, or 12% for 40% position falsifying nodes. Position faking with dropping results in higher loss, namely about 20% respectively 32%.

Figure 6 contains the relative delivery ratio reduction for different network sizes, compared to the case without falsified position information. When malicious nodes do not drop packets, increasing network sizes continue to reduce the relative delivery ratio. With packet drops, we observe a maximum reduction at network sizes of $2500m * 2500m$. This is the result of two overlapping effects. On the one hand, with increased network size, the number of hops and thus the probability of encountering a malicious node increases. On the other hand, with sparse network density, the probability of unsuccessful delivery due to network partitioning increases anyway and leverages the effects of dropping. The latter effect is visualized in figure 7, where the overall delivery ratio is shown for different network area sizes.

3.3   Analysis of Reasons for Decreased Delivery Ratio

The decreased amount of successfully delivered messages in scenarios with position falsifying nodes has its origin in three different reasons of messages getting "lost" during their traversal of the ad hoc network. These three are, packet drops due to detection of routing loops, undelivered messages remaining in caches since no suitable next hop has been found and packets dropped by maliciously acting nodes. Obviously, the latter reason is only of importance in scenarios, where position faking nodes actually drop packets.

According to our assumptions made in section 2, one reason for the decreased ratio of successfully delivered messages is the higher amount of packet drops due to routing loops. Figure 8 shows the corresponding simulation results. As a general remark, larger network sizes result in higher number of intermediate hops and therefore in a higher probability for creation of routing loops. From the simulation results in figure 8, we see, packet drops resulting from detected routing loops do also occur, even if there is no falsified position information. This results from the combination of node mobility and packet caching as recovery strategy. In scenarios, where position falsifying nodes do not drop received packets, the amount of packets dropped due to routing loops is always higher. On the other hand, it is obvious that if position faking nodes do drop received packets, i.e. before they can get into routing loops, this value has to be inferior.

The simulation results for the second reason for decreased delivery ratio, the amount of packets remaining in the node's caches, is shown in figure 9. According to these simulation results, in most cases falsified position information does not cause an increased number of packets remaining in the caches. For scenarios without packet dropping by maliciously acting nodes, the results are quite close

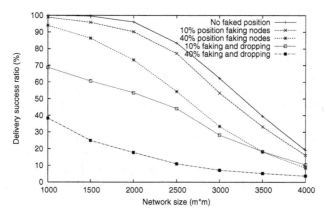

Fig. 7. Percentage of successfully delivered messages for different network sizes

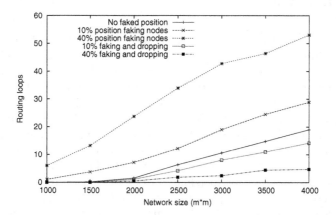

Fig. 8. Number of drops due to routing loops

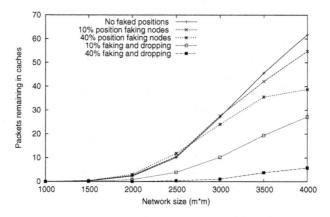

Fig. 9. Number of undelivered messages remaining in caches

to those of simulations without false position information. The increasing difference for larger network areas is caused by the increasing amount of packet drops in routing loops. And again, in case, maliciously acting nodes do drop packets, this effect can be neglected.

As an overall conclusion of this analysis, we retain the following. Depending on the behavior of position faking nodes, the following effects are responsible for the decreased ratio of successfully delivered messages. If the falsifying nodes do not drop packets, the main reason are packet drops resulting from detected routing loops. Their number is higher than the reduction of packets remaining in the routing caches compared to the case without faked positions. If the falsifying nodes maliciously do drop packets, the dropping itself is the dominant effect. Improvements regarding both other effects are only the result of less packets remaining in the network after those drops.

4   Related Work

The possibility of using *geographic* routing for mobile ad hoc networks has been investigated intensely. Especially the vision of ad hoc routing in vehicular networks was a stimulus for geographic routing research. This is due to the particular characteristics of such networks on the one hand and the necessity of geographic data distribution for the envisioned applications on the other hand.

Among the proposed packet forwarding schemes based on the individual node position, some main categories can be identified [6]. One of these comprises the *greedy routing* approaches. All greedy approaches have in common that the next hop node of a packet has to be closer to the destination's position than the current node. In case multiple neighbors satisfy this criterion, several selection strategies were proposed. The greedy-only method selects the neighbor with the smallest Euclidean distance to the destination. In contrast, MFR (most forward progress within radius [12]) projects the positions of the suitable neighbors onto a straight line stretched across the current node's position and the destination's position. Then, the neighbor with the most "progress" on that line is chosen. Other greedy methods select the next hop randomly or by the minimal distance to the current node (NFP [13]) in order to save sending power. Obviously, all greedy methods are stuck if there is no neighbor closer to the destination's position. If packets shall not be lost at such a point, a recovery strategy must be introduced. The perimeter routing in GPSR [14] is one possibility, caching the packet until a suitable neighbor appears is another [10].

A completely different geographic routing category uses *restricted directional flooding* [6]. For example, the LAR (Location aided routing) protocol by Ko and Vaidya [15] defines a rectangular region with the sender's position as one edge, and the destination's position as the diagonal opposed edge. Within that region, the packet is flooded. DREAM [16] acts very similarly, but uses a conus-shaped flooding region.

A third category of geographic routing applies hierarchical mechanisms. Terminodes [17], for instance, introduces two levels of routing. In a small region of several hops, a proactive routing is used, whereas larger distances are traversed by a special greedy method.

For vehicular ad hoc networks, geographic routing is particularly appropriate. Car-to-car networks show high node mobility and contain potentially large numbers of nodes. Geographic routing is able to address these challenges better than topology-based protocols [18]. One reason is that topology-based protocols like DSR or AODV need to find and maintain routes, which is not necessary for geographic routing. The matter of position determination is not a critical issue in vehicular ad hoc networks, due to the increasing number of cars being equipped with GPS receivers, which is mostly used in navigation systems.

Kim, Lee and Helmy have conducted examinations on the impact of location inaccuracies on geographic routing [19]. They defined a scheme to classify localization errors and ran simulations with relative location errors ranging from $0m$ to $50m$. They simulated using GPSR, with and without perimeter mode. Their results show some effects like routing loops that have also been observed during our work, under the assumption of malicious nodes.

Apart from these observations of localization errors and in contrast to routing functionality, there has been no work on security concerns specific to effects of falsified position data in geographic ad hoc routing.

5   Conclusion

Falsified position information in mobile ad hoc networks with geographic routing protocols results in serious network performance degradation. In this paper we have presented an analysis of local and global effects of falsified position information. Our simulation results show that the overall delivery ratio might decrease even up to approximately 30%, depending on the number of maliciously acting nodes and depending on whether the malicious nodes drop packets or not.

Furthermore, we analyzed the reasons for decreased delivery ratio, which again, depend on the forwarding behavior of malicious nodes. Whereas for scenarios without packet dropping by position faking nodes, drops resulting from routing loops are the main reason, in scenarios with packet dropping by position faking nodes, the dropping itself is the actual reason.

In current research, we develop methods to detect maliciously acting nodes, in order to lower the effects of faked position information. These methods comprise detection techniques and countermeasures, which are divided into single node and co-operative functions.

References

1. Franz, W., Wagner, C., Maihöfer, C., Hartenstein, H.: Fleetnet: Platform for inter-vehicle communications. In: Proc. 1st International Workshop on Intelligent Transportatin (WIT'04), Hamburg, Germany (2004)
2. CarTalk 2000. (http://www.cartalk2000.net)
3. US Vehicle Safety Communication Consortium. (http://http://www-nrd.nhtsa.dot.gov/pdf/nrd-12/CAMP3/pages/VSCC.htm)

4. Network on Wheels. (http://www.informatik.uni-mannheim.de/pi4/lib/projects/NoW/links.html)
5. Global Systems For Telematics. (http://www.gstproject.org/)
6. Mauve, M., Widmer, J., Hartenstein, H.: A survey on position-based routing in mobile ad-hoc networks (2001)
7. Kargl, F., Schlott, S., Weber, M., Klenk, A., Geiss, A.: Securing ad hoc routing protocols. In: Proceedings of 30th Euromicro Conference, Rennes, France (2004)
8. Kargl, F., Gei, A., Schlott, S., Weber, M.: Secure dynamic source routing. In: Proceedings of the 38th Hawaii International Conference on System Sciences (HICSS-38), Hilton Waikoloa Village, HA (2005)
9. Hubaux, J.P., Čapkun, S., Luo, J.: The security and privacy of smart vehicles. IEEE Security and Privacy **4** (2004) 49–55
10. Maihöfer, C., Eberhardt, R., Schoch, E.: CGGC: Cached Greedy Geocast. In: Proc. 2nd Intl. Conference Wired/Wireless Internet Communications (WWIC 2004). Volume 2957 of Lecture Notes in Computer Science., Frankfurt (Oder), Germany, Springer Verlag (2004)
11. Saha, A.K., Johnson, D.B.: Modeling mobility for vehicular ad-hoc networks. In: VANET '04: Proceedings of the first ACM workshop on Vehicular ad hoc networks, ACM Press (2004) 91–92
12. Takagi, H., Kleinrock, L.: Optimal transmission ranges for randomly distributed packet radio terminals. IEEE Transactions on Communications **32** (1984) 246–257
13. Hou, T.C., Li, V.: Transmission range control in multihop packet radio networks. IEEE Transactions on Communications **34** (1986) 38–44
14. Karp, B., Kung, H.: Greedy perimeter stateless routing for wireless networks. In: Proceedings of the Sixth ACM/IEEE International Conference on Mobile Computing and Networking (MobiCom 2000), Boston, USA (2000) 243–254
15. Ko, Y., Vaidya, N.: Location-aided routing (lar) in mobile ad hoc networks. In: Proceedings of the Fourth ACM/IEEE International Conference on Mobile Computing and Networking (MobiCom 1998). (1998) 66–75
16. Basagni, S., Chlamtac, I., Syrotiuk, V.R., Woodward, B.A.: A distance routing effect algorithm for mobility (DREAM). In: Proceedings of the ACM/IEEE International Conference on Mobile Computing and Networking (MobiCom), Dallas, USA, ACM Press (1998) 76–84
17. Blazevic, L., Giordano, S., Boudec, J.L.: Self organized terminode routing. Technical Report DSC/2000/040, Swiss Federal Institute of Technology (2000)
18. Füssler, H., Mauve, M., Hartenstein, H., Käsemann, M., Vollmer, D.: A comparison of routing strategies for vehicular ad hoc networks. Technical Report TR-3-2002, Department of Computer Science, University of Mannheim (2002)
19. Kim, Y., Lee, J.J., Helmy, A.: Impact of location inconsistencies on geographic routing in wireless networks. In: MSWIM '03: Proceedings of the 6th ACM international workshop on Modeling analysis and simulation of wireless and mobile systems, ACM Press (2003) 124–127

Provable Security of On-Demand
Distance Vector Routing
in Wireless Ad Hoc Networks

Gergely Ács, Levente Buttyán, and István Vajda

Laboratory of Cryptography and Systems Security (CrySyS),
Department of Telecommunications,
Budapest University of Technology and Economics, Hungary
{acs, buttyan, vajda}@crysys.hu

Abstract. In this paper, we propose a framework for the security analysis of on-demand, distance vector routing protocols for ad hoc networks, such as AODV, SAODV, and ARAN. The proposed approach is an adaptation of the simulation paradigm that is used extensively for the analysis of cryptographic algorithms and protocols, and it provides a rigorous method for proving that a given routing protocol is secure. We demonstrate the approach by representing known and new attacks on SAODV in our framework, and by proving that ARAN is secure in our model.

1 Introduction

Routing is a fundamental networking function, which makes it an ideal starting point for attacks aiming at disabling the operation of an ad hoc network. Therefore, securing routing is of paramount importance. Several "secure" routing protocols for ad hoc networks have been proposed in the academic literature (see [7] for a good overview), but their security have been analyzed by informal means only. In [3] and in [1], we show that flaws in routing protocols can be very subtle (leading to very sophisticated attacks), and therefore, they are very difficult to discover by informal reasoning. In [1], we propose a systematic approach based on a mathematical framework, in which the security of on-demand source routing protocols (e.g., DSR [8], Ariadne [6], and SRP [11]) can be analyzed rigorously. In this paper, we extend that approach to on-demand, distance vector protocols (e.g., AODV [12], SAODV [14] and ARAN [13]).

We must emphasize that by secure routing we mean the security of the route discovery part of the routing protocol only. In other words, we are not concerned with the problem of misbehaving nodes that do not forward data packets either for selfish or for malicious reasons. There are many attacks that aim at paralyzing the entire network by denial of service (e.g., rushing attack) or subverting the neighbor discovery mechanism (e.g., wormhole attack). In our notion these are not against the route discovery process primarily, and thus, we are not concerned with them in the rest of the paper.

R. Molva, G. Tsudik, and D. Westhoff (Eds.): ESAS 2005, LNCS 3813, pp. 113–127, 2005.
© Springer-Verlag Berlin Heidelberg 2005

Rather, we focus on the problem of how to maintain the "correctness" of the routing information stored in the routing tables of the honest nodes in the presence of an adversary. We will define precisely what we mean by the "correctness" of routing table entries later in this paper.

Our mathematical framework is based on the simulation paradigm that has been successfully used to analyze the security of various cryptographic algorithms and protocols (see parts V and VI of [9] for an overview). In this approach, one constructs a real-world model that describes the real operation of the system, and an ideal-world model that captures what the system wants to achieve in terms of security. Then, in order to prove the security of the system, one proves that the outputs of the two models are indistinguishable (statistically or computationally). In [1], we apply this approach to source routing protocols, where the output of the models are sets of routes returned by the routing protocol in route reply messages. In case of distance vector routing, however, no routes are returned explicitly in the route reply messages. Hence, the main novelty of this paper is that here, the output of the models is the state of the system, which is represented by the content of the routing tables of the honest nodes.

The rest of the paper is organized as follows. In Section 2, we introduce our mathematical framework, which includes a precise definition of a "correct" system state, and based on that, a definition of routing security. Then, in Section 3, we illustrate the concepts introduced in Section 2 by representing known and new attacks on SAODV in our framework. In Section 4, we demonstrate the usefulness of our approach by formally proving that ARAN is secure in our model. Finally, we report on some related work in Section 5, and conclude the paper in Section 6.

2   Model

2.1   Static Representation of the System

Network model: We model the ad hoc network as an undirected labelled graph $G(V, E)$, where V is the set of vertices and E is the set of edges. Each vertex represents a node, and there is an edge between two vertices if and only if there is a radio link between the corresponding nodes. We assume that the radio links are symmetric, and that is why the graph is undirected.

We assume that the nodes use authenticated identifiers (e.g., public keys, symmetric keys) during neighbor discovery and in the routing protocol. We denote the set of identifiers by L, and we label each vertex v of G with the identifiers used by the node corresponding to v. We assume that honest (non-corrupted) nodes use a single identifier that is unique in the network, whereas corrupted nodes may use multiple compromised identifiers (see attacker model below). We represent the assignment of identifiers to the nodes by a labelling function $\mathcal{L} : V \to 2^L$, which returns for each vertex v the set of labels assigned to v. As we mentioned above, if v corresponds to a non-corrupted node, then $\mathcal{L}(v)$ is a singleton, and $\mathcal{L}(v) \not\subseteq \mathcal{L}(v')$ holds for any other vertex v'.

We also assign cost values to the nodes and to the radio links that may be interpreted as (minimum) processing and transmission costs, respectively, and may be used to compute routing metrics. The assignment of cost values is represented by two functions $\mathcal{C}_{node} : V \to \mathbb{R}$ and $\mathcal{C}_{link} : E \to \mathbb{R}$. Quite naturally, $\mathcal{C}_{node}(v)$ will represent the cost assigned to the node that corresponds to vertex v, and $\mathcal{C}_{link}(e)$ will represent the cost assigned to the link that corresponds to edge e. In the following, we will omit the indices *node* and *link* of $\mathcal{C}$ when the type of the argument unambiguously determines which of the two functions is used in a given context. An example for a typical cost assignment is the following: $\mathcal{C}(v) = 1$ for all $v \in V$, and $\mathcal{C}(e) = 0$ for all $e \in E$, which leads to the widely used hop count metric, where the cost of a route is equal to the number of intermediate nodes on the route.

Adversary model: We assume that the adversary is not all powerful, but it launches its attacks from corrupted nodes that it controls and that have similar communication capabilities as regular nodes. We denote the vertices that correspond to corrupted nodes by V^*. In addition, we assume that the adversary compromised some identifiers, by which we mean that the adversary compromised the cryptographic keys that are used to authenticate those identifiers. We denote the set of compromised identifiers by L^*. We further assume that the adversary distributed all compromised identifiers to all corrupted nodes, and hence, we have $\mathcal{L}(v) = L^*$ for all $v \in V^*$. Using the notation introduced in [6], the adversary described above is an Active-y-x adversary, where $x = |V^*|$ and $y = |L^*|$. In addition, we assume that the adversary is static in the sense that it does not corrupt more nodes and compromise more identifiers during the operation of the system.

Since neighboring corrupted nodes can communicate with each other in an unrestricted manner (e.g., by sending encrypted messages), they can appear as a single node (under all the compromised identifiers) to the other nodes. Hence, without loss of generality, we assume that corrupted nodes are not neighbors in G; if they were, we could merge them into a single corrupted node that would inherit all the neighbors of the original nodes.

Configuration: A configuration is a five tuple $(G(V, E), V^*, \mathcal{L}, \mathcal{C}_{node}, \mathcal{C}_{link})$ that consists of the network graph, the set of corrupted nodes, the labelling function, and the cost functions.

2.2   System States and Correctness

The state of the system is represented by the routing tables of the non-corrupted nodes. We assume that an entry of the routing table of a given node v contains the following three fields: the identifier of the target node, the identifier of the next hop towards the target, and the cost value that represents the believed cost of the route to the given target via the given next hop. Without loss of generality, we assume that the routing metric is such that routes with lower cost values are preferred.

Consequently, the state of the system in our model will be represented by a set $Q \subset (V \setminus V^*) \times L \times L \times \mathbb{R}$ of quadruples such that for any $(v, \ell_{tar}, \ell_{nxt}, c)$ and $(v', \ell'_{tar}, \ell'_{nxt}, c')$ in Q, $v = v'$ and $\ell_{tar} = \ell'_{tar}$ and $\ell_{nxt} = \ell'_{nxt}$ implies $c = c'$. The quadruple $(v, \ell_{tar}, \ell_{nxt}, c)$ in Q represents an entry in v's routing table with target identifier ℓ_{tar}, next hop identifier ℓ_{nxt}, and believed route cost c. The ensemble of quadruples that have v as their first element represent the entire routing table of v, and the ensemble of all quadruples in Q represent the ensemble of the routing tables of the non-corrupted nodes (i.e., the state of the system). Note that we allow that a node's routing table contains multiple entries for the same target, but the next hops should be different.

We define a correct state as follows:

Definition 1 (Correct state). *A state Q is correct if for every $(v, \ell_{tar}, \ell_{nxt}, c) \in Q$, there exists a sequence $v_1, v_2, \ldots, v_p$ of vertices in V such that $(v_i, v_{i+1}) \in E$ for all $1 \le i < p$, and*

- $v_1 = v$,
- $\ell_{tar} \in \mathcal{L}(v_p)$,
- $\ell_{nxt} \in \mathcal{L}(v_2)$*, and*
- $\sum_{i=2}^{p-1} C_{node}(v_i) + \sum_{i=1}^{p-1} C_{link}(v_i, v_{i+1}) \le c$.

Intuitively, the system is in a correct state if all the routing table entries of the non-corrupted nodes are correct in the sense that if v has an entry for target ℓ_{tar} with next hop ℓ_{nxt} and cost c, then indeed there exists a route in the network that

- starts from node v
- ends at a node that uses the identifier ℓ_{tar}
- passes through a neighbor of v that uses identifier ℓ_{nxt}, and
- has a cost that is smaller than or equal to c.

The requirement on the believed cost of the route (last point above) deserves some explanation. First of all, recall the assumption that routes with a lower cost are preferred. It is, therefore, natural to assume that the adversary wants to make routes appearing less costly than they are. This means that if node v believes that there exists a route between itself and target ℓ_{tar} (passing through neighbor ℓ_{nxt}) with a cost c, while in reality, there exist only routes between them with a cost higher than c, then the system should certainly be considered to be in an incorrect state (i.e., under attack). On the other hand, allowing the existence of routes with a smaller cost does not have any harm (under the assumption that the adversary has no incentive to increase the believed costs corresponding to the routes), and it makes the definition of the correct state less demanding. This has a particular importance in case of protocols that use one-way hash chains to protect hop count values (e.g., SAODV and alike), since in those protocols, the adversary can always increase the hop count by hashing the current hash chain element further. However, this ability of the attacker should rather be viewed as a *tolerable imperfection* of the system than a flaw in those protocols.

2.3   Dynamic Representation of the System

The simulation paradigm: The main idea of the simulation paradigm is to define two models: a real-world model that represents the behavior of the real system and an ideal-world model that describes how the system should work ideally. In both models, there is an adversary, whose behavior is not constrained apart from requiring it to run in time polynomial in the security parameter (e.g., size of the cryptographic keys used by the cryptographic primitives). This allows us to consider *any* feasible attacks, and makes the approach very general. Although the adversary is not constrained, the construction of the ideal-world model ensures that all of its attacks are unsuccessful against the ideal-world system. In other words, the ideal-world system is secure by construction.

Once the models are defined, the goal is to prove that for any real-world adversary, there exist an ideal-world adversary that can achieve essentially the same effects in the ideal-world model as those achieved by the real-world adversary in the real-world model (i.e., the ideal-world adversary can simulate the real-world adversary). A successful proof means that no attacks can be successful in the real-world model (or more precisely attacks can be successful only with negligible probability), since otherwise, an attack would be successful in the ideal-world model too, which is impossible by definition.

Real-world model: The real-world model that corresponds to a configuration $conf = (G(V, E), V^*, \mathcal{L}, \mathcal{C}_{node}, \mathcal{C}_{link})$ and adversary $\mathcal{A}$ is denoted by $sys^{real}_{conf,\mathcal{A}}$, and it is illustrated on the left hand side of Figure 1. $sys^{real}_{conf,\mathcal{A}}$ consists of a set $\{M_1, \ldots, M_n, A_1, \ldots, A_m, H, C\}$ of interacting Turing machines, where the interaction is realized via common tapes. Each M_i represents a non-corrupted device that corresponds to a vertex in $V \setminus V^*$, and each A_j represents a corrupted vertex in V^*. H is an abstraction of higher-layer protocols run by the honest parties, and C models the radio links represented by the edges in E. All machines are probabilistic.

We describe the operation of the real-world model only briefly, since it is essentially the same as the operation of the model described in [1]. Each machine is initialized with some input data (e.g., identifiers of neighbors, cryptographic keys, etc.), which determines its initial state. In addition, the machines also receive some random input (the coin flips to be used during the operation). Once the machines have been initialized, the computation begins. The machines operate in a reactive manner, which means that they need to be activated in order to perform some computation. When a machine is activated, it reads the content of its input tapes, processes the received data, updates its internal state, writes some output on its output tapes, and goes back to sleep. The machines are activated in rounds by a hypothetic scheduler (not illustrated in Figure 1). The order of activation is not important, apart from the requirement that C must be activated at the end of the round.

Machine C is intended to model the broadcast nature of radio communications. Its task is to read the content of the output tape of each machine M_i and A_j and copy it on the input tapes of *all* the neighboring machines, where the neighbor relationship is determined by the configuration *conf*. Machine H models higher-layer protocols (i.e., protocols above the routing protocol) and the end-users of the non-corrupted devices. H can initiate a route discovery process at any machine M_i by placing a request on tape req_i. A response to this request is eventually returned via tape res_i. Machines M_i ($1 \leq i \leq n$) represent the non-corrupted nodes, which belong to the vertices in $V \setminus V^*$. M_i communicates with the other protocol machines via its output tape out_i and its input tape in_i, and its operation is essentially defined by the routing algorithm.

Finally, machines A_j ($1 \leq j \leq m$) represent the corrupted nodes, which belong to the vertices in V^*. Regarding its communication capabilities, A_j is identical to any machine M_i. However, A_j may not follow the routing protocol faithfully. In addition, A_j may send out-of-band requests to H by writing on ext_j by which it can instruct the honest parties to initiate route discovery processes. Here, we make the restriction that the adversary triggers a route discovery only between non-corrupted nodes. Moreover, we restrict each A_j to write on ext_j only once, at the very beginning of the computation (i.e., before receiving any messages from other machines). This essentially means that we assume that the adversary is *non-adaptive*; it cannot initiate new route discoveries as a function of previously observed messages. Note, however, that each A_j can write multiple requests on ext_j, which means that we allow several parallel runs of the routing protocol.

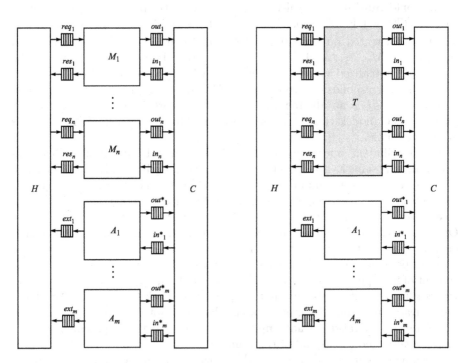

Fig. 1. The real-world model $sys^{real}_{conf,\mathcal{A}}$ (left hand side) and the ideal-world model $sys^{ideal}_{conf,\mathcal{A}}$ (right hand side)

The computation ends when H reaches one of its final states. This happens when H receives a response to each of the requests that it placed on the tapes req_i ($1 \le i \le n$), where a response can also be a time-out. The output of $sys^{\text{real}}_{conf,\mathcal{A}}$ is the ensemble of the routing tables of the non-corrupted nodes, which is a set of quadruples as defined above in Subsection 2.2. We denote the output by $Out^{\text{real}}_{conf,\mathcal{A}}(r)$, where r is the random input of the model. In addition, $Out^{\text{real}}_{conf,\mathcal{A}}$ will denote the random variable describing $Out^{\text{real}}_{conf,\mathcal{A}}(r)$ when r is chosen uniformly at random.

Ideal-world model: The ideal-world model that corresponds to a configuration $conf = (G(V,E), V^*, \mathcal{L}, \mathcal{C}_{node}, \mathcal{C}_{link})$ and adversary $\mathcal{A}$ is denoted by $sys^{\text{ideal}}_{conf,\mathcal{A}}$, and it is illustrated on the right hand side of Figure 1. One can see that the ideal-world model is similar to the real-world model; the main difference is that machines M_i ($1 \le i \le n$) are replaced with a new machine called T. The operation of the ideal-world model is very similar to the real-world model, therefore, we do not detail it here. We focus only on the operation of the new machine T.

In effect, machine T emulates the behavior of the machines M_i ($1 \le i \le n$), with the difference that T is initialized with $conf$, and hence, it can detect when the system gets into an incorrect state. When this happens, T records that the system has been in an incorrect state, but the computation continues as if nothing wrong had happened.

Similar to the real-world model, the computation ends, when H reaches one if its terminal states, which happens when H receives a response to each of the requests that it placed on the tapes req_i ($1 \le i \le n$), where a response can also be a time-out. The output of the ideal-world model is either the ensemble of the routing tables if T has not recorded an incorrect state during the computation, or a special symbol that indicates that an incorrect state has been encountered. The output is denoted by $Out^{\text{ideal}}_{conf,\mathcal{A}}(r)$. Moreover, $Out^{\text{real}}_{conf,\mathcal{A}}$ denotes the random variable describing $Out^{\text{real}}_{conf,\mathcal{A}}(r)$ when r is chosen uniformly at random.

2.4   Definition of Security

Based on the model introduced in the previous subsections, we define routing security formally as follows:

Definition 2. *(Statistical security) A routing protocol is said to be statistically secure if, for any configuration conf and any real-world adversary $\mathcal{A}$, there exists an ideal-world adversary $\mathcal{A}'$, such that $Out^{\text{real}}_{conf,\mathcal{A}}$ is statistically indistinguishable[1] from $Out^{\text{ideal}}_{conf,\mathcal{A}'}$.*

The intuitive meaning of the definition above is that if a routing protocol is statistically secure, then any system using this routing protocol gets into an incorrect

---

[1] Two random variables are statistically indistinguishable if the L_1 distance of their distributions is negligibly small.

state only with negligible probability. This negligible probability is related to the fact that the adversary can always forge the cryptographic primitives (e.g., generate a valid digital signature) with a very small probability.

3   Insecurity of SAODV

SAODV [14] is a "secure" variant of the Ad hoc On-demand Distance Vector (AODV) [12] routing protocol. In the following, we briefly overview the operation of SAODV, and we show that, in fact, it is not secure in our model.

3.1   Operation of SAODV

The operation of SAODV is similar to that of AODV, but it uses cryptographic extensions to provide integrity of routing messages and to prevent the manipulation of the hop count information. Conceptually, SAODV routing messages (i.e., route requests and route replies) have a non-mutable and a mutable part. The non-mutable part includes, among other fields, the node sequence numbers, the addresses of the source and the destination, and a request identifier, while the mutable part contains the hop count information. Different mechanisms are used to protect the different parts.

The non-mutable part is protected by the digital signature of the originator of the message (i.e., the source or the destination of the route discovery). This ensures that the non-mutable fields cannot be changed by an adversary without the change being detected by the non-corrupted nodes.

In order to prevent the manipulation of the hop count information, the authors propose to use hash chains. When a node originates a routing message (i.e., a route reply or a route request), it first sets the HopCount field to 0, and the MaxHopCount field to the TimeToLive value. Then, it generates a random number *seed*, and puts it in the Hash field of the routing message. After that, it calculates the TopHash field by hashing *seed* iteratively MaxHopCount times. The MaxHopCount and the TopHash fields belong to the non-mutable part of the message, while the HopCount and the Hash fields are mutable. Every node receiving a routing message hashes the value of the Hash field (MaxHopCount − HopCount) times, and verifies whether the result matches the value of the TopHash field. Then, before rebroadcasting a route reply or forwarding a route request, the node increases the value of the HopCount field by one, and updates the Hash field by hashing its value once.

The rationale behind using the above hash chaining mechanism is that given the values of the Hash, the TopHash, and the MaxHopCount fields, anyone can verify the value of the HopCount field. On the other hand, preceding hash values cannot be computed starting from the value in the Hash field due to the one-way property of the hash function. This ensures that an adversary cannot decrease the hop count, and thus, cannot make a route appearing shorter than it really is. However, as we will see later (and as pointed out by the authors of SAODV themselves), this latter statement does not hold in general, because a corrupted node that happens to be on a route between the source and the destination may pass on the routing message without increasing the value of the HopCount field and without updating the value of the Hash field.

3.2   Simple Attacks Against SAODV

According to our definition of security, a routing protocol is secure if it ensures that incorrect entries in the routing tables of the non-corrupted nodes can be generated only with negligible probability. In case of SAODV, a node v creates an entry in its routing table for a target ℓ_{tar} only if it receives a fresh enough routing message that carries a valid digital signature of ℓ_{tar}. The fact that this routing message arrived to v means that there must be a route between v and a node that uses the identifier ℓ_{tar}, since otherwise, the message cannot reach v. However, SAODV cannot guarantee that the next hop and the hop count information in the newly created routing table entry is correct. This is illustrated by the following two examples.

Attack 1: Let us consider the configuration illustrated in Figure 2. Since SAODV uses the hop count as the routing metric, we set the node cost to 1 for every node and the link cost to 0 for every link. Let us assume that the node labelled by S starts a route discovery towards the node labelled by T. When the route request message reaches the corrupted node labelled by Z, it does not increase the hop count and does not update the hash value in the message. Therefore, when this route request is eventually received by the node labelled by T, it will create an entry $(S, B, 1)$ in its routing table. In addition, this entry will not be overwritten when the other route request message arrives through the node labelled by C, since that request will have a hop count of 2. This means that the system ends up in an incorrect state, because there is not any route in this network that starts at the node labelled by T, passes through the node labelled by B, ends at the node labelled by S, and has a cost less than or equal to 1.

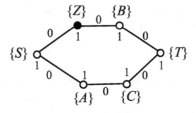

Fig. 2. A configuration where the adversary can achieve that the node labelled by T creates in its routing table an entry with an incorrect cost value when SAODV is used

We note that this weakness of SAODV has already been known by its authors (see Subsection 5.3.5 of [14]). Our purpose with this example is simply to illustrate how an attack that exploits the weakness can be represented within our framework.

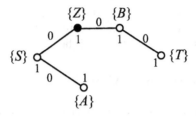

Fig. 3. A configuration where the adversary can achieve that the node labelled by S creates in its routing table an entry for target T with an incorrect next hop A when SAODV is used

Attack 2: Let us now consider the configuration illustrated in Figure 3. Let us assume again that the source is the node labelled by S and the destination is the node labelled by T. Furthermore, let us assume that a route request message reached the destination, and it returned an appropriate route reply. When this reply reaches the corrupted node labelled by Z, it forwards it to the node labelled by S in the name of A. Therefore, the node labelled by S will create a routing table entry $(T, A, 2)$. Note, however, that there is no route at all from the node labelled by S to the node labelled by T that passes through the node labelled by A. In other words, the system ends up in an incorrect state again. To the best of our knowledge, this weakness of SAODV has not been published yet.

4   Security of ARAN

ARAN (Authenticated Routing for Ad hoc Networks) is another secure, distance vector routing protocol for ad hoc networks proposed in [13]. In this section, we briefly overview its operation, and we prove that it is secure in our model.

4.1   Operation of ARAN

Just like SAODV, ARAN as well uses public key cryptography to ensure the integrity of routing messages. Initially, a source node S begins a route discovery process by broadcasting a route request message:

$$(\mathsf{RREQ}, T, cert_S, N_S, t, Sig_S)$$

where RREQ means that this is a route request, S and T are the identifiers of the source and the destination, respectively, N_S is a nonce generated by S, t is the current time-stamp, $cert_S$ is the public-key certificate of the source, and Sig_S is the signature of the source on all of these elements. N_S is a monotonically increasing value that, together with t and S, uniquely identifies the message, and it is used to detect and discard duplicates of the same request (and reply).

Later, as the request is propagated in the network, intermediate nodes also sign it. Hence, the request has the following form in general:

$$(\mathsf{RREQ}, T, cert_S, N_S, t, Sig_S, Sig_A, cert_A)$$

where A is the identifier of the intermediate node that has just re-broadcast the request. When a neighbor of A, say B, receives this route request, then it verifies both signatures, and the freshness of the nonce. If the verification is successful, then B sets an entry in its routing table with S as target, and A as next hop. Then, B removes the certificate and the signature of A, signs the request, appends its own certificate to it, and rebroadcasts the following message:

$$(\mathsf{RREQ}, T, cert_S, N_S, t, Sig_S, Sig_B, cert_B)$$

When destination T receives the first route request that belongs to this route discovery, it performs verifications and updates it routing table in a similar manner as it is done by the intermediate nodes. Then, it sends a route reply message to S. The route reply is propagated back on the reverse of the discovered route as a unicast message. The route reply sent by T has the following form:

$$(\mathsf{RREP}, S, cert_T, N_S, t, Sig_T)$$

where RREP means that this is a route reply, N_S and t are the nonce and the time-stamp obtained from the request, S is the identifier of the source, $cert_T$ is the public-key certificate of T, and Sig_T is the signature of T on all of these elements.

Similar to the route request, the route reply is signed by intermediate nodes too. Hence, the general form of the route reply is the following:

$$(\mathsf{RREP}, S, cert_T, N_S, t, Sig_T, Sig_B, cert_B)$$

where B is the identifier of the node that has just passed the reply on.

A node A that receives the route reply verifies both signatures in it, and if they are valid, then it forwards the reply to the neighbor node from which it has received the corresponding route request previously. However, before doing that, A will remove the certificate and the signature of B, and put its own certificate and signature in the message:

$$(\mathsf{RREP}, S, cert_T, N_S, t, Sig_T, Sig_A, cert_A)$$

In addition, A also sets an entry in its routing table for target T with B as the next hop.

As it can be seen from the description, ARAN does not use hop counts as a routing metric. Instead, the nodes update their routing tables using the information obtained from the routing messages that arrive first; any later message that belongs to the same route discovery is discarded. This means that ARAN may not necessarily discover the shortest paths in the network, but rather, it discovers the quickest ones. In effect, ARAN uses the message propagation delay (i.e., physical time) as a path length metric.

4.2   Security Proof

Theorem 1. *ARAN is a secure ad hoc routing protocol in our model, if the signature scheme is secure against chosen message attacks.*

Proof. Since ARAN uses the message propagation delay as the routing metric, we will assume that the node cost values in our model represent minimum message processing delays (at the nodes), and the link cost values represent minimum message transmission delays (on the links). In addition, we make the pessimistic assumption that the adversary's message processing delay is 0, which means that $\mathcal{C}_{node}(v) = 0$ for all $v \in V^*$.

In order to be compliant with our framework, we also assume that each routing table entry explicitly contains a routing metric value too. In our case, this metric value is the time that was needed for the routing message that triggered the creation of this entry to get from the originator of the message to the node that created this entry. Although these times are not represented explicitly in ARAN routing table entries, representing them in the model does not weaken our results in any way. In particular, exactly the same routing table entries are created in our model as in ARAN with respect to the target and the next hop identifiers.

In order to prove that ARAN is secure, one has to find the appropriate ideal-world adversary $\mathcal{A}'$ for any real-world adversary $\mathcal{A}$ such that Definition 2 is satisfied. Due to the constructions of our models, a natural candidate is $\mathcal{A}' = \mathcal{A}$, since in that case, the steps of the real-world and the ideal-world models are exactly the same (for the same random input, of course). If no incorrect state is encountered during the computation in the ideal-world model, then not only the steps, but the outputs of the two models will be the same too. On the other hand, if an incorrect state occurs in the ideal-world model, then the outputs of the models will be different, since the ideal-world model will output a special symbol. Hence, Definition 2 is satisfied, if an incorrect state can only be encountered with negligible probability. We will show that indeed this is the case for ARAN.

Getting into an incorrect state means that one of the non-corrupted nodes, say v, sets an incorrect entry in its routing table. Let this incorrect entry be $(\ell_{tar}, \ell_{nxt}, c)$. Since v is non-corrupted, it sets this entry only if it received a routing message that has been signed by ℓ_{tar} as originator and ℓ_{nxt} as previous hop, v has a neighbor that uses identifier ℓ_{nxt}, and it took time c for the message to get from its originator to v. Now, $(\ell_{tar}, \ell_{nxt}, c)$ can be incorrect for one of the following three reasons:

1. There is no route from v to a node that uses ℓ_{tar}.
2. There are routes from v to a node that uses ℓ_{tar}, but none of them go through any neighbor of v that uses ℓ_{nxt}.
3. There are routes from v to a node that uses ℓ_{tar} going through a neighbor of v using ℓ_{nxt}, but each of them has a cost higher than c.

In case 1, if the signature of ℓ_{tar} in the routing message is not forged, then the very fact that v received the message proves that there is a route between v and a node that uses ℓ_{tar} (since otherwise the message could not reach v). Hence, case 1 is possible only if the signature of ℓ_{tar} is forged, and this has negligible probability if the signature scheme is secure.

In case 2, if the signature of ℓ_{nxt} in the routing message is not forged, then a neighbor of v, say v', that uses ℓ_{nxt} has indeed seen and signed the message. Now, the same reasoning can be used for v' as in case 1 for v: if the signature of ℓ_{tar} in the routing message is not forged, then the fact that v' received the message proves that there is a route between v' and a node that uses ℓ_{tar}, and hence, there is a route between v and a node that uses ℓ_{tar} that goes through v' (since v' is a neighbor of v). This means that case 2 is possible only if the signature of ℓ_{tar} or ℓ_{nxt}, or both are forged, and this has negligible probability.

Finally, in case 3, let R be the set of existing routes that start at v, end at a node that uses ℓ_{tar}, and go through a neighbor of v using ℓ_{nxt}. Moreover, let c' be the minimum of the costs of the routes in R. By assumption, $c' > c$. If the signatures of ℓ_{tar} and ℓ_{nxt} in the routing message received by v are not forged, then the message must have taken one of the routes in R. However, it could not reach v in time $c < c'$, since the node and link costs represent the minimum message processing and transmission delays at the nodes and on the links. In other words, the adversary cannot speed up the transmissions on the links and the processing at the non-corrupted nodes. Hence, case 3 is possible only if either ℓ_{tar} or ℓ_{nxt}, or both are forged, which can happen only with negligible probability. ∎

5 Related Work

There are several proposals for secure ad hoc routing protocols (see [7] for a recent overview). However, most of these proposals come with an informal security analysis with all the pitfalls of informal security arguments. Another set of related papers deal with provable security for cryptographic algorithms and protocols (see Parts V and VI of [9] for a survey of the field) and with the application of formal methods for the security analysis of cryptographic protocols (see [4] for an overview of the main approaches). However, these papers are not concerned with ad hoc routing protocols. There exist only a few papers where formal techniques are proposed for the verification of the security of ad hoc routing protocols; we briefly overview them here.

In [15], the authors propose a formal model for ad hoc routing protocols that is similar to the strand spaces model [5], which has been developed for the formal verification of key exchange protocols. Routing security is defined in terms of a safety and a liveness property. The liveness property requires that it is possible to discover routes, while the safety property requires that discovered routes do not contain corrupted nodes. In contrast to this, our definition of security admits routes that pass through corrupted nodes, because it seems to be impossible to guarantee that discovered routes do not contain any corrupted node, given that corrupted nodes can behave correctly and follow the routing protocol faithfully.

Another approach, presented in [10], is based on a formal method, called CPAL-ES, which uses a weakest precondition logic to reason about security protocols. Unfortunately, the work presented in [10] is very much centered around the analysis of SRP [11], and it is not general enough. We must also mention that in [11], SRP has been analyzed by its authors using BAN logic [2]. However, BAN logic has never been intended for the analysis of routing protocols, and there is no easy way to represent the requirements of routing security in it. In addition, a basic assumption of BAN logic is that the protocol participants are trustworthy, which does not hold in the typical case that we are interested in, namely, when there are corrupted nodes in the network controlled by the adversary that may not follow the routing protocol faithfully.

Finally, in [3] and [1], we have developed and applied an approach based on the simulation paradigm for on-demand source routing protocols for ad hoc networks. The framework proposed in this paper is essentially the adaptation of that approach to on-demand, distance vector routing protocols.

6   Conclusion

In this paper, we proposed an approach for the security analysis of on-demand, distance vector routing protocols for ad hoc networks, such as AODV, SAODV, and ARAN. The proposed approach is based on the simulation paradigm that is used extensively for the analysis of cryptographic algorithms and protocols, and it provides a rigorous method for proving that a given routing protocol is secure. We demonstrated the approach by representing two attacks on SAODV in our framework, and by proving that ARAN is secure in our model.

An important message of this paper is that flaws (leading to attacks) in ad hoc routing protocols can be very subtle, and hard to discover by informal reasoning. Another important message is that it is possible to adopt sound analysis techniques known from the cryptographic literature, and to use them in the context of ad hoc routing protocols.

In our future work, we intend to automate (at least partially) the process of the security analysis of ad hoc routing protocols. For this purpose, we will identify an appropriate formal framework, e.g., one based on model checking. Furthermore, our current definition of a correct state is not strict enough, because it does not consider that an adversary might have an interest in increasing the cost of a route passing through it (perhaps, to get rid of the traffic). Thus, we intend to extend the definition of a correct routing table entry by requiring an appropriate upper bound on the believed cost of the route.

Acknowledgements

The work presented in this paper has partially been supported by the Hungarian Scientific Research Fund (T046664). The second author has been further supported by IKMA and by the Hungarian Ministry of Education (BÖ2003/70).

References

[1] G. Ács, L. Buttyán, and I. Vajda. Provably Secure On-demand Source Routing in Mobile Ad Hoc Networks. Technical Report, Budapest University of Technology and Economics, March 2005. Available on-line at: http://www.hit.bme.hu/~buttyan/publications.html

[2] M. Burrows, M. Abadi, and R. Needham. A logic of authentication. *ACM Transactions on Computer Systems*, 8(1):18–36, February 1990.

[3] L. Buttyán and I. Vajda. Towards provable security for ad hoc routing protocols. In *Proceedings of the ACM Workshop on Security in Ad Hoc and Sensor Networks (SASN)*, October 2004.

[4] R. Focardi and R. Gorrieri (eds). *Foundations of Security Analysis and Design.* LNCS 2171, Springer-Verlag, 2000.

[5] J. Guttman. Security goals: packet trajectories and strand spaces. In *Foundations of Security Analysis and Design*, edited by R. Focardi and R. Gorrieri, Springer LNCS 2171, 2000.

[6] Y.-C. Hu, A. Perrig, and D. Johnson. Ariadne: A secure on-demonad routing protocol for ad hoc networks. In *Proceedings of the ACM Conference on Mobile Computing and Networking (Mobicom)*, 2002.

[7] Y.-C. Hu and A. Perrig. A survey of secure wireless ad hoc routing. *IEEE Security and Privacy Magazine*, 2(3):28–39, May/June 2004.

[8] D. Johnson and D. Maltz. Dynamic source routing in ad hoc wireless networks. In *Mobile Computing*, edited by Tomasz Imielinski and Hank Korth, Chapter 5, pages 153–181. Kluwer Academic Publisher, 1996.

[9] W. Mao. *Modern Cryptography: Theory and Practice.* Prentice Hall PTR, 2004.

[10] J. Marshall. An Analysis of the Secure Routing Protocol for mobile ad hoc network route discovery: using intuitive reasoning and formal verification to identify flaws. MSc thesis, Department of Computer Science, Florida State University, April 2003.

[11] P. Papadimitratos and Z. Haas. Secure routing for mobile ad hoc networks. In *Proceedings of SCS Communication Networks and Distributed Systems Modelling Simulation Conference (CNDS)*, 2002.

[12] C. Perkins and E. Royer. Ad hoc on-demand distance vector routing. In *Proceedings of the IEEE Workshop on Mobile Computing Systems and Applications*, pp. 90-100, February 1999.

[13] K. Sanzgiri, B. Dahill, B. Levine, C. Shields, and E. Belding-Royer. A secure routing protocol for ad hoc networks. In *Proceedings of the International Conference on Network Protocols (ICNP)*, 2002.

[14] M. G. Zapata and N. Asokan. Securing ad hoc routing protocols. *Proceedings of the ACM Workshop on Wireless Security (WiSe)*, 2002.

[15] S. Yang and J. Baras. Modeling vulnerabilities of ad hoc routing protocols. In *Proceedings of the ACM Workshop on Security of Ad Hoc and Sensor Networks*, October 2003.

Statistical Wormhole Detection
in Sensor Networks

Levente Buttyán, László Dóra, and István Vajda

Laboratory of Cryptography and System Security (CrySyS),
Department of Telecommunications,
Budapest University of Technology and Economics, Hungary
{buttyan, laszlo.dora, vajda}@crysys.hu

Abstract. In this paper, we propose two mechanisms for wormhole detection in wireless sensor networks. The proposed mechanisms are based on hypothesis testing and they provide probabilistic results. The first mechanism, called the Neighbor Number Test (NNT), detects the increase in the number of the neighbors of the sensors, which is due to the new links created by the wormhole in the network. The second mechanism, called the All Distances Test (ADT), detects the decrease of the lengths of the shortest paths between all pairs of sensors, which is due to the shortcut links created by the wormhole in the network. Both mechanisms assume that the sensors send their neighbor list to the base station, and it is the base station that runs the algorithms on the network graph that is reconstructed from the received neighborhood information. We describe these mechanisms and investigate their performance by means of simulation.

1 Introduction

Sensor networks [1] consist of a large number of sensors that monitor the environment, and a few base stations that collect the sensor readings. The sensors are usually battery powered and limited in computing and communication resources, while the base stations are considered to be more powerful. In order to reduce the overall energy consumption of the sensors, it is conceived that the sensors send their readings to the base station via multiple wireless hops. Hence, in a sensor network, the sensor nodes are responsible not only for the monitoring of the environment, but also for forwarding data packets towards the base station on behalf of other sensors.

In order to implement the above described operating principle, the sensors need to be aware of their neighbors, and they must also be able to find routes to the base station. An adversary may take advantage of this, and may try to control the routes and to monitor the data packets that are sent along these routes [6]. One way to achieve this is to set up a *wormhole* in the network. A wormhole is a dedicated connection between two physical locations. The adversary installs a radio transceiver at each end of the connection, and it sends and re-transmits every data packet received at one end of the wormhole at the

R. Molva, G. Tsudik, and D. Westhoff (Eds.): ESAS 2005, LNCS 3813, pp. 128–141, 2005.
© Springer-Verlag Berlin Heidelberg 2005

other end of it. Practically, this means that the adversary creates communication links between some pairs of sensors that would otherwise not be able to communicate directly with each other. In other words, the adversary modifies the topology of the network. If this is done carefully, the adversary may achieve that many sensors send their data packets to the base station via the wormhole. While the application of cryptographic mechanisms (e.g., encryption and message authentication) prevents an adversary from monitoring and modifying the information sent to the base station, cryptographic mechanisms do not solve every problem stemming from wormholes. The adversary can still mount denial of service type attacks, such as dropping packets (possibly selectively) that are transferred through the wormhole. In addition, the sensors which are close to the transceivers of the wormhole participate more in packet forwarding and they deplete their battery earlier. Therefore, in most of the applications, wormhole detection is an important requirement.

In this paper, we propose two mechanisms for wormhole detection in wireless sensor networks. The proposed mechanisms are based on hypothesis testing and they provide probabilistic results. The first mechanism, called the Neighbor Number Test (NNT), detects the increase in the number of the neighbors of the sensors, which is due to the new links created by the wormhole in the network. The second mechanism, called the All Distances Test (ADT), detects the decrease of the lengths of the shortest paths between all pairs of sensors, which is due to the shortcut links created by the wormhole in the network. Both mechanisms assume that the sensors send their neighbor list to the base station, and it is the base station that runs the algorithms on the network graph that is reconstructed from the received neighborhood information. The main advantage of the proposed mechanisms is that they do not require special hardware in the sensors, directional antennas, tight clock synchronization, or distance measurements between the nodes. The only requirement is that the sensor nodes can determine who their neighbors are, and they can send this information to the base station in a secure way. The rest of the paper is organized as follows. In Section 2, we overview the state-of-the-art in the filed of wormhole detection. In Section 3, we present our approach by describing the operation of the two wormhole detection mechanisms that we propose. The effectiveness of the mechanisms is studied in Section 4, where we present and analyze our simulation results. Finally, in Section 5, we conclude the paper and sketch some possible future research directions.

2   Related Work

In [5], the authors propose two approaches for detecting wormholes in wireless ad hoc networks, where sensors are allowed to move during the communication. The first approach is called *geographical packet leashes*, and it requires the nodes to be aware of their own location and to maintain loosely synchronized clocks. Every time when a node A sends a packet to its neighbor B, it puts its location and the time of sending into the header of the packet. When the packet is received by

B, it compares the time of reception to the time of sending, and calculates the maximum distance between A and B using the difference between their locations and the distance that they could move away between sending and receiving the packet. If the estimated distance is longer than the possible maximum radio range then B rejects the communication with A.

The other approach is called *temporal packet leashes*, and it avoids using any special hardware for localization, but it requires tightly synchronized clocks. Every time when a node A sends a packet to its neighbor B, it puts an authenticated time stamp into the header. When B receives the packet, it calculates the possible maximum distance between A and B from the difference between the time of sending and the time of receiving of the packet, and assuming that the packet travels with the speed of light. If the resulting distance is too large, then this indicates a wormhole. This procedure relies on the fact that going through the wormhole means covering a longer distance than the normal distance between neighboring nodes, and this longer distance can be precisely measured due to the tightly synchronized clocks.

The disadvantage of the above approaches is that they require either location information of each node or tight clock synchronization between the nodes, and these requirements cannot always be satisfied in sensor networks.

In [3], another approach is proposed to estimate the real physical distance between two communicating nodes, which does not require location information or clock synchronization at all. That approach is based on an authenticated *distance-bounding* protocol, called MAD. The distance-bounding phase of MAD consist of several rounds, and in each round, each node sends a one bit challenge to the other node to which the other node responds with a one bit response immediately. Each node locally measures the time between sending out the challenge and receiving the response, and based on the measured times, it estimates its distance to the other node, assuming that messages travel with the speed of light. In order for this to work, the nodes must be able to measure local timings with nanosecond precision, which is possible with today's hardware. In addition, it is crucial that the response is sent immediately after receiving the challenge. This, however, may not be possible using standard hardware. The main problem is that typical wireless medium access control protocols introduce random delays between the time at which the application sends a message and the time at which that message is really transmitted via the radio interface. Therefore, this approach also requires special hardware in the sensor nodes and special medium access control protocols.

Another wormhole detection approach that uses the node's location information is proposed in [7]. However, as opposed to the geographical leash approach proposed in [5], here only a small fraction of the nodes need to be equipped with a GPS receiver. These special nodes are called *guards* and it is also assumed that the guards have a larger radio range (denoted by R) than the other nodes. The guards broadcast their positions in their one hop neighborhood. Two nodes consider each other neighbor only if they hear a threshold number of common guards. The nodes use the location information broadcast by the guards to detect wormholes based on the following two principles: (i) since any guard heard by a node must lie within a range of radius R around the node, a node cannot hear two guards that are $2R$ apart from each other; and (ii) since the messages sent by the guards are authenticated and protected against replay, a node cannot receive the same message twice from the same guard. It is shown in [7] that based on these principles, wormholes can be detected with probability close to one. However, the disadvantage of this approach is that the guards are distinguished nodes in the network that differ from the regular nodes.

In [4], the authors propose a wormhole detection approach that assumes that the nodes know from which direction they got a packet. The intuitive idea behind this approach is that if there is no wormhole in the system, then the following must be true: if one node sends a packet in a given direction, then its neighbor will hear that packet from the opposite direction. However, if there is a wormhole in the system, then the above statement is not always true (depending on the placement of the wormhole), and thus, the wormhole becomes detectable. Unfortunately, it has a significant probability that the wormhole is there, but it is not caught. In order to address this problem, the authors worked out two algorithms in which the nodes involve their neighbors during the communication to help to discover the wormhole. The main disadvantage of this approach is that it requires directional antennas, which are usually not available in sensor networks.

In [8], a centralized wormhole detection technique is proposed, which uses inaccurate distance estimations between neighboring nodes. The main idea of the proposed technique is to reconstruct a virtual layout of the network and identify inconsistencies in it. For this reason, the connectivity information and the inaccurately estimated distances between the neighbors are fed into a multidimensional scaling (MDS) algorithm, which tries to determine a virtual position for every node in such a way that the constraints induced by the connectivity and the distance estimation data are respected. Since the distances are estimated inaccurately, the algorithm has a certain level of freedom in "stretching" the nodes within the error bounds of the distance estimation. If the estimated distance between two nodes connected by a wormhole are much larger than the nodes' communication range, then the wormhole is detected immediately. Hence, the adversary must falsify the distance estimation and arrange that the estimated distances between the nodes affected by the wormhole become credible. However, this will result in a distortion in the virtual layout constructed by the MDS algorithm; in particular, the layout will be contracted between the affected nodes. By visualizing the virtual layout or by computing appropriate indicator values, the distortion can be detected and the wormhole can be located.

3   Our Approach

Compared to the above described approaches, our approach neither requires special hardware and directional antennas in the nodes, nor tight clock synchronization and distance measurements. We only assume that the sensor nodes can determine who their neighbors are, and they can send this information to the base station(s). Based on the received neighborhood information, the base station(s) can detect the presence of wormholes probabilistically using hypothesis testing. In this section, we propose two specific mechanisms for this purpose; we will evaluate the effectiveness of the proposed mechanisms in Section 4.

3.1   System Assumptions

We assume that the system consists of a large number of sensor nodes and a few base stations placed on a two dimensional surface. We assume that the base stations have no resource limitations, and they can run complex algorithms. We assume that the sensors have a fixed radio range r, and two sensors are neighbors, if they reside in the radio range of each other. We assume that the sensors run some neighbor discovery protocol, and they can determine who their neighbors are. We also assume that the sensors send their neighborhood information to the closest base station regularly in a secure way. By security we mean confidentiality, integrity, and authenticity; in other words, we assume that the adversary cannot observe and change the neighborhood information sent to the base stations by the sensors, neither can it spoof sensors and fabricate false neighborhood updates. This can be ensured by using cryptographic techniques that we will not detail in this paper. Note that the neighborhood information can be piggy-backed on regular data packets. In addition, as sensor networks tend to be rather static, sending only the changes in the neighborhood since the last update would reduce the overhead significantly. The base stations can pool the received neighborhood information together, and based on that, they can reconstruct the graph of the sensor network. We assume that the node density is high enough so that the network is always connected.

We assume that the adversary can set up a wormhole in the system. The wormhole is a dedicated connection between two physical locations. There are radio transceivers installed at both ends of the wormhole, and packets that are received at one end can be sent to and re-transmitted at the other end. In this way, the adversary can achieve that nodes that otherwise do not reside in each other's radio range can still hear each other and establish a neighbor relationship (i.e., they can run the neighbor discovery protocol). This means that the adversary can introduce new, otherwise non-existing links in the network graph that is constructed by the base stations based on the received neighborhood information.

The wormhole is characterized by the distance between the two locations that it connects and the radio ranges of its transceivers. We assume that the receiving and the sending ranges of both transceivers are the same, and we will call this range the *radius* of the wormhole. The radius of the wormhole is not necessarily equal to the radio range of the sensors.

In principle, the adversary can drop packets carrying neighborhood information that are sent to the base stations via the wormhole. However, consistently missing neighborhood updates can be detected by the base stations and they indicate that the system is under attack. Therefore, we assume that the adversary does not drop the neighborhood updates. In addition, by the assumptions made earlier, it cannot alter or fabricate them either.

3.2   Neighbor Number Test (NNT)

Our first detection mechanism is based on the fact that by introducing new links into the network graph, the adversary increases the number of neighbors of the nodes within its radius. This is illustrated in Figure 1. The thick circle in the figure is the radio range of the sensor node A. Its real neighbors are N_i within the radio range of sensor node A. The two other circles show the radius of the wormhole. The nodes at the further end of the wormhole that are labelled with W_i are the neighbors that are due to the existence of the wormhole. These sensors are outside of the radio range of A, and they would not be its neighbors if there was no wormhole.

If the distribution of the placement of the nodes is given, then it is possible to compute the hypothetical distribution of the number of neighbors. Then, the base stations can use statistical tests to decide if the network graph constructed from the neighborhood information that is received from the sensors corresponds to

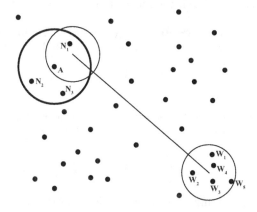

Fig. 1. The wormhole increases the number of neighbors of the nodes in its radius

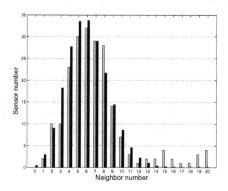

Fig. 2. Hypothetical (dark) and real (light) distributions of the number of neighbors

this hypothetical distribution. In order to illustrate this idea, let us consider the example depicted in Figure 2, where the dark bars correspond to the hypothetical distribution of the number of neighbors, and the light bars show the actual distribution in the network graph reconstructed from the sensors' neighborhood updates. One can see that the probability of higher neighbor numbers (15-20) is increased with respect to the hypothetical distribution, and the idea of the proposed mechanism is to detect this increase by using statistical tests.

Based on the above observations, the NNT algorithm is given as follows:

1. The base station computes the expected histogram of the neighbor numbers using the hypothetical distribution of the number of neighbors.
2. The base station collects the neighborhood updates from the sensors, constructs the network graph, and computes the histogram of the real neighbor numbers in the graph.
3. The base station compares the two histograms with the χ^2–test.
4. If the computed χ^2 number is larger than a preset threshold that corresponds to a given significance level, then a wormhole is indicated.

Computing the parameters for the χ^2–test. Assuming that the sensors are placed uniformly at random on the plane, the probability of two nodes being neighbors is

$$q = \frac{r^2 \cdot \pi}{T}$$

where r is the radio range of the sensor nodes and T is the sphere of the area where the sensor network is deployed. The probability $p(k)$ of having exactly k neighbors is

$$p(k) = \binom{N}{k} \cdot q^k \cdot (1-q)^{N-k}$$

where $N+1$ is the total number of nodes in the network. Let us partition the set $\{0, 1, 2, \ldots\}$ into subsets $B_1, B_2, \cdots, B_m$, such that $e(i) = (N+1)\sum_{k \in B_i} p(k)$ be larger than 5 (a requirement needed by the χ^2–test [2]). The χ^2 number is then computed using the following formula:

$$\chi^2 = \sum_{\forall i} \frac{r(i) - e(i)}{e(i)}$$

where $r(i)$ is the real number of nodes with number of neighbors in B_i. If χ^2 is below the threshold that corresponds to a given significance level (this threshold can be looked up in published tables of χ^2 values), then the hypothesis is accepted, and no wormhole is indicated. Otherwise the hypothesis is rejected, and a wormhole is indicated.

3.3 All Distances Test (ADT)

Our second detection mechanism is based on the fact that the wormhole shortens the paths in the network, or more precisely, it distorts the distribution of the

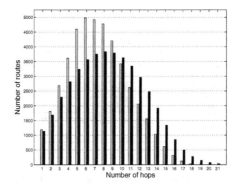

Fig. 3. Hypothetical (dark) and real (light) distributions of the length of the shortest paths between all pairs of nodes

length of the shortest paths between all pairs of nodes. This is illustrated by the example depicted in Figure 3, where the dark bars represent the hypothetical distribution of the length of the shortest paths and the light bars represent the real distribution. As it can be seen, the two distributions are different, and in the real distribution, shorter paths are more likely than in the hypothetical one. The idea is to detect this difference with statistical tests.

The ADT algorithm is very similar to the NNT algorithm:

1. The base station computes the histogram of the length of the shortest paths between all pairs of nodes in the hypothetical case when there is no wormhole in the system using the knowledge of the distribution of the node placement.
2. The base station collects the neighborhood information from the sensors, and computes the histogram of the length of the shortest paths in the real network.
3. The base station compares the two histograms with the χ^2–test.
4. If the computed χ^2 number is larger than a preset threshold that corresponds to a given significance level, then a wormhole is indicated.

Computing the parameters for the χ^2–test. In this case, we were not able to derive a close formula that describes the hypothetical distribution of the length of the shortest paths. Instead, we propose to estimate that distribution by randomly placing nodes on the plane according to the distribution of the node placement, and compute the lengths of the shortest paths between all pairs of nodes in the resulting graph. We propose to repeat the experience many times and average the normalized histograms obtained in these experiences. Once the hypothetical distribution is estimated in this way, the χ^2–test can be used in a similar way as we described in Subsection 3.2.

4   Simulation Environment

In order to evaluate the effectiveness of the proposed mechanisms, we built a simulator that places 300 sensor nodes uniformly at random on a 500 m × 500 m

Table 1. Simulation parameters

Number of nodes	300
Extent of territory	500 m × 500 m
Number of simulation runs	100
Radio range of sensor nodes	40 m, 47 m, 54 m, 60 m, 65 m, 70 m
Radio range of the wormhole	16 m, 50 m
Distance between the affected areas at the two end wormhole	20 m, 50 m, 100 m, 200 m, 300 m, 400 m

flat area with one base station in the middle, and it also places a wormhole randomly in the same area. The simulator permits us to set three parameters: the radio range of the sensors, the radius of the wormhole, and the distance between the affected areas at the two ends of the wormhole.

We chose two extreme values for the radio range of the sensor nodes: 40 m and 70 m. The expected neighbor number is 5.9 in the 40 m case, and 18.5 in the 70 m case. Then, we split up the range between 5.9 and 18.5 evenly into 5 pieces to get the six radio range values that we used in our simulations (see Table 1).

We set the radius of the wormhole to 16 m or to 50 m (see Table 1). These two values have been selected in such a way that the number of nodes affected by the wormhole differs significantly in the two cases. When the radius of the wormhole is 16 m, one node is affected (falls in the wormhole's range) on both ends of the wormhole on average, whereas when the radius of the wormhole is 50 m, 9.4 nodes are affected on both ends on average.

Finally, we varied the distance between the affected areas at the two ends of the wormhole between 20 m and 400 m (see Table 1).

A given combination of the possible parameter values define a test case. For each test case we run 100 simulations and averaged the results. For each radio range setting, we first determined the rate of the false positive alarms (i.e., the percentage of the simulation runs where the algorithms indicate a wormhole when there is no wormhole in the system). Then, we placed wormholes with different parameters in the system and determined the accuracy of both of our detection mechanisms (i.e., the percentage of simulation runs where the wormhole is detected when there is indeed a wormhole in the system). The results are presented in the following subsections.

4.1   Results of the Neighbor Number Test (NNT)

The results of the NNT algorithm are shown on Figures 4 and 5. Figure 4(a) shows the accuracy of the detection as a function of the radio range of the sensors when the radius of the wormhole is 50 m. As it can be seen, the detection accuracy decreases as the sensors' radio range increases. The reason is that in the case of larger radio ranges, the sensors have more real neighbors, and therefore, the increase in the number of neighbors caused by the wormhole becomes less significant, and consequently, more difficult to detect. We can also observe that the detection accuracy is better when the areas affected by the wormhole are

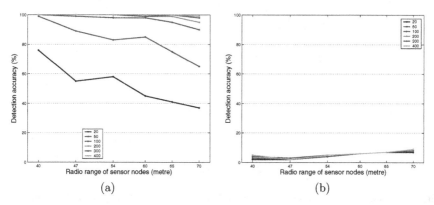

Fig. 4. Detection accuracy plotted against the radio range of the sensor nodes. The different curves belong to different distances between the areas affected by the wormhole with a radius of 50 m (a) and 16 m (b).

more distant from each other, although increasing this distance above 100 m has no real influence on the results. In fact, if the distance between the affected areas is smaller than the radio range of the sensors, then it is possible that two affected nodes that do not belong to the same affected areas are already real neighbors, and therefore, the wormhole does not create a new link between them. In other words, the larger the distance between the affected areas is, the higher the probability is that the wormhole introduces new links into the graph, and by doing so it increases the number of neighbors of the affected nodes.

Figure 4(b) shows the accuracy of the detection as a function of the radio range of the sensors when the radius of the wormhole is 16 m. It is clear from the figure that the NNT algorithm does not work in this case, as the accuracy of the detection is unacceptably low. The huge difference between the performance in the 50 m case and that in the 16 m case can be explained with the large difference in the number of the affected nodes in the two cases. As we described earlier, when the radius of the wormhole is 16 m, on average one node is affected at both ends on the wormhole. Hence, practically, such a wormhole creates a single new link in the graph, which is extremely difficult detect with statistical techniques. On the other hand, as the average number of affected nodes is around 10 at both ends of the wormhole when the radius is 50 m, the number of new links introduced in the graph is around 100. More importantly, around 20 nodes out of the total of 300 have around 10 more neighbors due to the wormhole, and this can be detected by the NNT algorithm.

Figure 5 shows the percentage of the false positive alarms as a function of the radio range of the sensors. As it can be seen, the NNT algorithm performs quite well regarding the false positive alarms. Indeed, the percentage of the false positive alarms is determined by the selected significance level of the χ^2-test, which in our case was 0.025.

In summary, the NNT algorithm detects the wormhole reasonably well if the radius of the wormhole is comparable to or larger than the radio range of the sensors, but it performs very badly if the radius of the wormhole is small. We

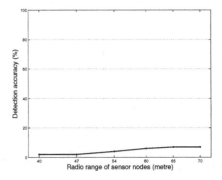

Fig. 5. Percentage of false positive wormhole detections plotted against the radio range of sensor nodes

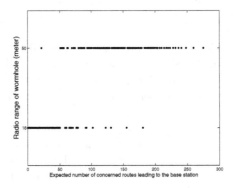

Fig. 6. The effect of the wormhole on the number of the controlled shortest paths plotted against the radius of the wormhole

note, however, that a smaller wormhole radius has smaller effect on the system in terms of the number of sensors that send measurement data to the base station through the wormhole. In order to illustrate this, we constructed the minimum spanning tree rooted at the base station, and counted the number of shortest paths between the base station and the sensors that contain a link created by the wormhole. The result is shown in Figure 6. As it can be seen, when the radius of the wormhole is 16 m, the number of concerned paths is between 0 and 50, whereas in the case of a 50 m radius, the number of concerned paths is between 100 and 200. Thus, the adversary can monitor the measurements of more sensors when the radius of the wormhole is larger, but in that case, it can also be detected more accurately by the NNT algorithm.

4.2   Results of the All Distances Test (ADT)

The results of the ADT algorithm are shown on Figures 7 and 8. Figure 7(a) shows the accuracy of the detection as a function of the sensors' radio range when the radius of the wormhole is 50 m, whereas Figure 7(b) shows the same

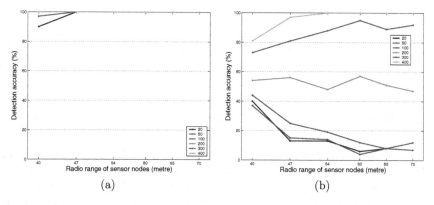

Fig. 7. Detection accuracy plotted against the radio range of the sensor nodes. The different curves belong to different distances between the areas affected by the wormhole with a radius of 50 m (a) and 16 m (b).

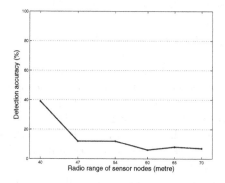

Fig. 8. Percentage of false positive wormhole detections plotted against the radio range of the sensor nodes

when the radius of the wormhole is 16 m. Similar to the NNT algorithm, the ADT algorithm performs better when the radius of the wormhole is larger. However, unlike the NNT algorithm, the ADT algorithm is not completely unusable in the case when the radius of the wormhole is 16 m. Rather, its performance depends on the distance between the areas affected by the wormhole: the higher this distance is, the more accurate the detection is. Moreover, when the distance between the affected areas is 400 m, the accuracy is close to 100% . The explanation for this is quite obvious: a longer wormhole reduces the length of the shortest paths between more distant nodes, and thus overall, it represents a larger decrease in the average length of the shortest paths between all pairs of nodes.

Regarding the percentage of the false positive alarms (Figure 8), the ADT algorithm performs quite well except for small radio ranges.

One may have expected that the detection accuracy of the ADT algorithm is independent of the radius of the wormhole. The rationale would be that no

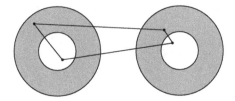

Fig. 9. Shortest paths are longer when the radius of the wormhole is smaller

matter how many new links are created by the wormhole, the important thing is that it creates shortcuts in the graph which reduce the lengths of the shortest paths between the sensors. However, this intuition is fallacious: shortest paths are indeed longer if the radius of the wormhole is smaller. As an illustration, let us consider Figure 9. The upper two nodes are directly connected if the radius is larger, whereas they are three hops away if the radius is small. This difference may seem to be small, but note that many shortest paths may use the wormhole and this two hop difference appears in each of them.

5   Conclusion and Future Work

In this paper, we have studied the problem of wormhole detection in wireless sensor networks. We proposed two mechanisms for wormhole detection that are based on hypothesis testing, and that provide probabilistic results. The first mechanism, called the Neighbor Number Test (NNT), detects the increase in the number of the neighbors of the sensors, which is due to the new links created by the wormhole in the network. The second mechanism, called the All Distances Test (ADT), detects the decrease of the lengths of the shortest paths between all pairs of sensors, which is due to the shortcut links created by the wormhole in the network. Both mechanisms assume that the sensors send their neighbor list to the base station, and it is the base station that runs the algorithms on the network graph constructed from the received neighborhood information.

We investigated the effectiveness of the two proposed mechanisms by means of simulation. Our results show that both mechanisms can detect the wormhole with high accuracy when the radius of the wormhole is comparable to the radio range of the sensors. In addition, the ADT algorithm performs better than the NNT algorithm when the radius of the wormhole is small (compared to the radio range of the sensors). In terms of false alarms, both algorithms perform reasonably well.

One disadvantage of the mechanisms that we proposed in this paper is that they detect only the presence of a wormhole, but they do not pinpoint its location. While detection is certainly the first thing that one needs to do, localization of the wormhole afterwards is also necessary for a successful defense. In the future, we intend to study if the statistical approach proposed in this paper can be extended to provide also wormhole localization services.

In this paper, we addressed the problem of wormhole detection in a static setting. In the future, we intend to extend our results to the dynamic case, when the wormhole is not present in the system from the beginning, but it is established by the adversary during the operation of the network. To some extent, detecting a dynamic wormhole is easier than detecting a static one: if the base station detects that two sensors that previously were many hops away from each other become neighbors, then it is reasonable to assume that a wormhole has just been established between them. On the other hand, such a detection scheme would require the sensors to provide neighborhood information to the base station continuously; a prohibitive price in sensor networks. Therefore, we are interested in the trade-offs between the overhead, the speed, and the accuracy of the detection.

Acknowledgement

The work presented in this paper has partially been supported by the Hungarian Scientific Research Fund (T046664). The first author has been further supported by IKMA and by the Hungarian Ministry of Education (BÖ2003/70).

References

1. I.F. Akyildiz, W. Su, Y. Sankarasubramaniam, E. Cayirci. Wireless sensor networks: a survey. *Computer Networks* 38:393-422, 2002.
2. I.N. Bronstein, K.A. Semendjajew, G. Musiol, and H. Muehlig. *Handbook of Mathematics*, Springer, 2004.
3. S. Čapkun, L. Buttyán, and J.-P. Hubaux. SECTOR: secure tracking of node encounters in multi-hop wireless networks. In *Proceedings of the ACM Workshop on Security in Ad Hoc and Sensor Networks (SASN)*, 2003.
4. L. Hu and D. Evans. Using directional antennas to prevent wormhole attacks. In *Proceedings of the IEEE Symposium on Network and Distributed System Security (NDSS)*, 2004.
5. Y. Hu, A. Perrig, and D. Johnson. Packet leashes: a defense against wormhole attacks in wireless ad hoc networks. In *Proceedings of the IEEE Conference on Computer Communications (Infocom)*, 2003.
6. C. Karlof and D. Wagner. Secure routing in sensor networks: attacks and countermeasures. *Ad Hoc Networks* 1:293–315, 2003.
7. R. Poovendran and L. Lazos. A graph theoretic framework for preventing the wormhole attack in wireless ad hoc networks, to appear in *ACM Wireless Networks*.
8. W. Wang and B. Bhargava. Visualization of wormholes in sensor networks. In *Proceedings of the ACM Workshop on Wireless Security (WiSe)*, 2004.

RFID System with Fairness Within the Framework of Security and Privacy*

Jin Kwak[1], Keunwoo Rhee[1], Soohyun oh[2], Seungjoo Kim[1], and Dongho Won[1]

[1] Information Security Group, Sunkyunkwan University,
300 Cheoncheon-dong, Jangan-gu, Suwon, Gyeonggi-do, 440-746, Korea
{jkwak, kwrhee, dhwon}@dosan.skku.ac.kr, skim@ece.skku.ac.kr
http://www.security.re.kr
[2] Division of Computer Science, Hoseo University,
29-1. Sechul-ri, Baebang-myun, Asan, Chuncheongnam-do, 336-795, Korea
shoh@office.hoseo.ac.kr

Abstract. Radio Frequency Identification (RFID) systems are expected to be widely deployed in automated identification and supply-chain applications. Although RFID systems have several advantages, the technology may also create new threats to user privacy. In this paper, we propose the Fair RFID system. This involves improving the security and privacy of existing RFID systems while keeping in line with procedures already accepted by the industrial world. The proposed system enables the protection of users' privacy from unwanted scanning, and, when necessary, is conditionally traceable to the tag by authorized administrators.

Keywords: Fairness, RFID, security, privacy, uncheckable, traceability.

1 Introduction

The Radio Frequency Identification (RFID) system is a technology that recognizes and manages the tag through the Radio Frequency (RF) signal. The low-cost RFID tag can be read, and information can be updated without physical contact. Therefore, RFID systems have become popular for automated identification and supply-chain applications. In addition, the RFID tag is expected to replace the barcode in supply-chain applications[4, 12, 17, 20, 22].

However, RFID system creates new problems such as the invasion of users' privacy including excessive information exposure. In particular, although the RFID tag enables more effective supply-chain management, it may also allow access to information regarding users credit information and purchase patterns without their agreement.

Thus, several methods and concepts for protecting the users' privacy have been proposed[7, 8, 9, 10, 14, 18, 19, 25, 26, 27]. However, some of these methods do not resolve privacy problems perfectly. (see [16] for more details.)

---

* This work was supported by the University IT Research Center Project funded by the Korean Ministry of Information and Communication.

R. Molva, G. Tsudik, and D. Westhoff (Eds.): ESAS 2005, LNCS 3813, pp. 142–152, 2005.
© Springer-Verlag Berlin Heidelberg 2005

The proposed system in this paper, can authenticate the tag without exposing its Unique IDentifier (UID) to the reader (or the back-end database), therefore, protecting users' privacy. If necessary, only authorized administrators can identify the UID of the tag. In the proposed system, when necessary, the UID of the tag can be tracked through the cooperation of authorized administrators.

The proposed system is applicable to user location based information services, such as missing children search, emergency call services, and so on. These services rely on the availability of user location information in order to provide specific targeted information. However, a user location is a sensitive piece of information and releasing it to random entities may create security and privacy issues. The user is entitled to protect their location information, keeping this information secret, or shared with trusted entities. Therefore, when required, only authorized entities should have access to this location information.

The subsequent sections of the paper are organized as follows. After shortly introducing the basic RFID system and components in Section 2, the security and privacy requirements for the proposed system are discussed in Section 3. In Section 4, the Fair RFID System is presented, enabling the protection of users' privacy, and, when necessary, is traceable to the tag by authorized administrators. The security properties of the proposed system are the described. Finally, in Section 6, this paper is concluded, with a discussion on possible future research directions.

2 RFID System Primer

RFID systems are basically composed of an RFID tag, RFID reader, and back-end database. The forward channel, i.e., the reader to the tag, is assumed to be broadcast with an RF signal to achieve long-range monitoring. However, the backward channel, i.e., the tag to the reader, in a relative secnse, is much weaker, enabling monitoring only by eavesdroppers within the tag's shorter operating range. In general, it is assumed that eavesdroppers can only monitor the forward channel undetected [3, 6, 21].

- **RFID tag** (transponder) includes UID (object-identifying data) [1]. Tags are generally composed of an IC chip and an antenna. The IC chip in the tag is used for data storage and logical operations, whereas the coiled antenna is used for communication between the reader and the tag. The RFID tag may either be an active or passive[2] tag. In this paper, the passive tag is focused on, in order to demonstrate hash operations.
- The **RFID reader** (scanner) is a device that transmits an RF signal to the tag, receives the information from the tag, and transmits such information to the back-end database. The reader may read and write data to the tag.

---

[1] In case of the location based service, the UID correspond to Social Security Number (SSN).

[2] The *Active tag* has a battery and actively sends the information to the reader. The *passive tag* must be inductively powered from the RF signal of the reader since RFID tags usually do not possess their own battery.

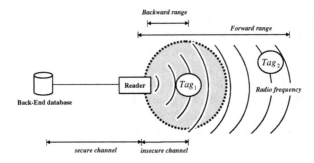

Fig. 1. Basic passive RFID System [21]

In general, readers are composed of the RF module, a control unit, and a coupling element to interrogate electronic tags via RF communication.

- The **Back-end database** is a data-processing system that stores related information[3] (e.g., product information, tracking logs, reader location, etc) with a particular tag.

3   Security and Privacy Requirements for the Propose System

To protect users' privacy, the UID of the tag should not be known to the legitimate reader and the back-end database; only the authentication of the legitimate tag should be provided. This section discusses the security requirements for the proposed system.

(1) Fairness

Let S be a RFID System, it can be said that S is a Fair RFID system if it guarantees a special agreed upon party under the proper circumstances envisaged by the policy to understand all UIDs encrypted using S, even without the users' consent and knowledge. If the encrypted ID is used in a Fair RFID system, only authorized administrators can retrieve the unique ID (UID) of the encrypted ID. That is to say, the RFID system should be able to provide not only user privacy but also traceability.

(2) Uncheckability

The recipient of response generated by the tag can verify that it is a valid response of the query, but cannot discover which tag made it.

(3) Anonymity

The recipient of a response generated by the tag can verify that it is a valid response of the query, but cannot decide whether two responses have been generated by the same tag. In this paper, the meaning of anonymity and that of "unlinkability" are the same.

---

[3] In case of the the location based services, user's telephone number, address, etc.

(4) Traceability

The UID of the tag should be encrypted and stored in *IMC* (UID Management Center), tracing of the UID, when necessary, should only be possible through the cooperation of authorized administrators.

4   Proposed RFID System with Fairness

This section proposes the system for the protection of users' privacy and, when necessary, the traceability of the UID of the tag. The readers are assumed to have a secure connection to the back-end database, although the tag and the reader communicate through an insecure channel. Fig.2 shows the proposed RFID System with Fairness.

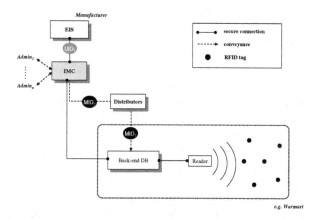

Fig. 2. Propose RFID System with Fairness

4.1   Initial Setup Phase

The initial setup phase of the proposed system is described, in the proposed system, to satisfy security requirements, and a cryptographic secret sharing method is adopted.

The proposes system consists of the *EIS (EPC Information Server)*[4], *IMC (UID Management Center)*[5], *RFID tag*, *RFID reader*, and *Back-end database*[6].

Fig.3 demonstrates the initial setup phase in the proposed system.

---

[4] It managed by the manufacturer. The *EIS* provides related information of products such as the term of validity, factory price, UID, and so on.

[5] The *IMC* generates *MID* (encrypted value of the tag's UID) of the tag. In this paper, for encryption of the *MID*, cryptographic secret sharing methods are adopted. [1, 2, 5, 11, 15, 23, 24])

[6] The retailer's retail system utilizing the information transmitted from the reader. It has sufficient computational ability.

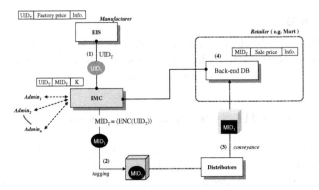

Fig. 3. Initial Setup phase

[Notations and Parameters]

- $h()$: collision-resistant hash-functions.
- $\|$: concatenation.
- $query_i$: query of R_i.
- UID_T : the unique identifier of the tag T
- MID_T : encrypted value of the tag's unique identifier[7]
- R_R : random number chosen by the reader.
- R_T : random number chosen by the tag.
- P : a set of participant Adm_is, $P = \{Adm_1, Adm_2, \cdots, Adm_n\}$.
- p : a large prime number, where $p > 2^{512}$.
- q : a prime number, where $q \mid p - 1$.
- g : an element over $\mathbb{Z}_p$, where $ord(g) = q$.
- Adm_i : administrators of IMC, they have secret sharing information of the key K.
- K : a master secret key used encryption of UID_T.
- y_{Adm} : a group public key of administrators.
- $ENC()$: a public key encryption scheme.

[Initial Setup]

1. The manufacturer issues a unique ID for each product by *EIS*. Then *EIS* sends the UID to the *IMC*.
2. The *IMC* encrypts the UID and then writes the encrypted value MID_T to the tag instead of the UID. The encryption process as follows.

   (The encryption of UID)

   ① Let there by n administrators. Every administrators ($Admin_i \mid i \in \{1, \cdots, n\}$) picks $r_i \in_R Z_q$ at random and broadcasts $y_i = g^{r_i} \bmod p$ to every other administrator in the set S_i.

   $$S_i = \{Adm_j \mid j \in \{1, \cdots, n\} \textbf{ AND } i \neq j\}$$

---

[7] In case of the location based service, *MID* is encrypted value of the user's SSN.

Each Adm_i picks $r_i \in_R Z_q$ at random and broadcasts $y_i = g^{r_i} \bmod p$ to all other administrators.

② To distribute r_i, each Adm_i randomly selects a polynomial f_i of degree $t - 1$ in Z_q such that $f_i(0) = r_i$, i.e.,

$$f_i(x) = r_i + a_{i,1}x + a_{i,2}x^2 + \cdots + a_{i,t-1}x^{t-1}$$

with $a_{i,1}, \cdots, a_{i,t-1} \in_R Z_q$, and sends $f_i(j) \bmod q$ to Adm_j in a secure manner ($\forall j \neq i$). Each Adm_i also broadcasts values

$$g^{a_{i,1}} \bmod p, \cdots, g^{a_{i,t-1}} \bmod p$$

③ From distributed $f_j(i)$ ($\forall j \neq i$), Adm_i checks whether, for each j,

$$g^{f_j(i)} = y_j \cdot (g^{a_{j,1}})^{i^1} \cdots (g^{a_{j,t-1}})^{i^{t-1}} \bmod p$$

④ Let $H = \{Adm_j \mid Adm_j$ is not detected to be cheating at step 3 $\}$. Every Adm_i computes the share

$$s_i = \sum_{j \in H} f_j(i)$$

secretly, and computes

$$y_{Adm} = \prod_{j \in H} y_j, \ g^{a_1} = \prod_{j \in H} g^{a_{j,1}}, \cdots, \ g^{a_{t-1}} = \prod_{j \in H} g^{a_{j,t-1}}$$

⑤ To encrypt each UID_i, Adm_1 picks $t_{i,1} \in_R Z_q$ and computes $g^{t_{i,1}} \bmod p$. Then, Adm_1 transmits the result to Adm_2.

⑥ The Adm_2 picks $t_{i,2} \in_R Z_q$ and computes $(g^{t_{i,1}})^{t_{i,2}} \bmod p$. Then, Adm_2 transmits the result to Adm_3.

⑦ The last participant $Adm_i(i \neq n)$ computes $g^{t_i} = (g^{t_{i,1},t_{i,2},\cdots})^{t_{i,n}} \bmod p$ and broadcasts the result to other Adm_n.

⑧ Through the cooperation of n Adm_i, the ciphertext of UID_i is generated as follow.

$$ENC(UID_i) = (g^{t_i}, \ (y_{Adm})^{t_i} \cdot SN_i \bmod p)$$

⑨ Each Adm_i stores $ENC(UID_i)$ in their IMC.

3. After step 2, the tag contains encrypted value instead of the UID- adhere to the product. Then the tagged products are transported to the retailer (e.g. Walmart).

4. When the tagged product arrives at the retailer, the products are managed by the retailer's systems.

4.2   Authentication Phase

The following steps present the authentication process of the proposed system. Fig.4 demonstrates the authentication phase of the proposed system.

1. RFID reader, which is connected with back-end database, broadcasts to the tags with a query and R_R.

· reader $\longrightarrow$ tag : $query$, R_R

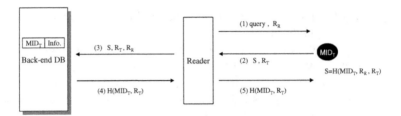

Fig. 4. Authentication protocol

2. When the tag received query and R_R, the tag generates random number R_T. Then the tag computes a response $S(= H(MID_T, R_T, R_R))$ and sends the S to the reader with R_T.

> · tag $\longrightarrow$ reader : S, R_T

3. The reader sends R_R to the back-end database with S and R_T received from the tag.

> · reader $\longrightarrow$ back-end database : S, R_T, R_R

4. For all MID_Ts in the back-end database, when the back-end database received data from the reader, the back-end database computes hash values. Then the back-end database compares it with S received from the reader to authenticate the tag.

> · back-end database : received $S \overset{?}{=}$ computed S

If the authentication is successful, the back-end database sends $H(MID_T, R_T)$ to the reader.

> · back-end database $\longrightarrow$ reader: $H(MID_T, R_T)$

5. The reader sends $H(MID_T, R_T)$ received from the back-end database to the tag.

> · reader $\longrightarrow$ tag: $H(MID_T, R_T)$

The tag computes $H(MID_T, R_T)$ and compares it with $H(MID_T, R_T)$ received from the reader to authenticate the back-end database. If the authentication is successful, the tag authenticates back-end database and the authentication session is successfully finished.

> · tag : computed $H(MID_T, R_T) \overset{?}{=}$ received $H(MID_T, R_T)$

4.3 Tracing Phase

If the IMC requires tracing the UID, he/she needs at least t $(n \geq t)$ shared information among n. This is because the UID is encrypted with the master secret key K, and K is shared between n administrators. Therefore, only authorized administrators can identify the UID of the tag.

If $X = \{Adm_1, Adm_2, \cdots, Adm_t\}$ is a qualified subset to recover the UID_i, they operate the recovery phase as follow.

1. Through the cooperation of every $Adm_i \in X$, the value $r_1 + r_2 + \cdots + r_t$ is recovered using polynomial interpolation [23].
2. Each $Adm_i \in X$ computes $K = (g^{t_i})^{r_1 + r_2 + \cdots + r_t}$ using stored g^{t_i}.
3. Each $Adm_i \in X$ computes the UID_i as follow

$$UID_i = ((y_{Adm})^{t_i} \cdot UID_i) \: / \: K$$

5 Security

In this section, the security of the proposed system is discussed. An unauthorized entity is assumed to eavesdrop on the RF signal between R and T. For security analysis, Ohkubo *et al.*'s methods of security proof are referred [14].

Definition 1. (*Fair RFID System*). The proposed system consists of tag T, reader R, back-end database DB, and management center IMC.

- T performs the following tasks:
  - after T receives *query* and R_R from R, T computes $S = H(MID_T, R_R, R_T)$ using received R_R and randomly chosen value R_T.
  - outputs S
- DB stores data set MID_T and operates as follows:
  - receives S, R_R, and R_T from R.
  - identifies T using MID_T, R_R, and R_T.

The property of *indistinguishability* is first discussed. The unauthorized entity Eve_{ind} is defined. Its purpose is to undermine the indistinguishability (i.e., attempt at tracking) property. Specifically, it attempts to distinguish between a truly random number and output S of T.

Definition 2. (*Uncheckability*). Unauthorized entity Eve_{ind} performs the following:

- access oracle T adaptively, transmits a query and R_R, then receives response (s, i).

$$s = S, \: i = R_T$$

- access oracle DB adaptively, transmits (s, i), and receives *response* or *error*.

$$response = H(MID_T, R_T)$$

Eve_{ind} requests for the problem with R_R, and $r \in_R \{0, 1\}$ is chosen randomly. If $r = 0$, Eve_{ind} is given (s', i) (where, $s' = S'$) response, which has not been given to Eve_{ind}. If $r = 1$, Eve_{ind} is given (a, i) response, where the value a is a truly random number. Finally, Eve_{ind} guesses r and outputs r'.

$$Advantage_{Eve_{ind}} = |Pr[r' \longleftarrow Eve_{ind}, r = r'] - 1/2|$$

In the proposed system, the advantage of any probabilistic polynomial-time of adversary Eve_{ind} is negligible. Therefore, the proposed system satisfies the property of *uncheckability*.

In addition, the output of each tag is different each time, since S is generated using the random value R_i. Therefore, it is impossible for an adversary to compute S without knowledge of the random values R_T, R_R and MID_T. Also, unauthorized tracking is impossible since a random answer is generated for each query.

Theorem 1. *Assuming functions H is random oracles, the proposed system is* ***uncheckable****.*

Secondly, the property of *unlinkability* is discussed. An adversary Eve_{link} is defined. Its purpose is to undermine the *unlinkability* property. In particular, it attempts to distinguish between (s, a) and (s, s'), where $s = S$, $s' = S'$ and a is a truly random number.

Definition 3 (*Unlinkability*). Adversary Eve_{link} performs the following:

- access oracle T adaptively, transmits a *query* and R_R, then receives a response (s, i).
$$s = S, \, i = R_T$$
- access oracle DB_1 and DB_2 adaptively, transmits (s, i), and receives *response* or *error*.
$$response = H(MID_T, R_T)$$

If Eve_{link} requests DB_1 and DB_2 for the problem with (ID_{DB_1}, ID_R) and (ID_{DB_2}, ID_R), $r \in_R \{0, 1\}$ is chosen randomly. If $r = 0$, Eve_{link} is provided a response $((s', i), (s'', i))$, which has not been given to A_{link}. If $r = 1$, Eve_{link} is provided a response $((s', i), (a, i))$, where value a is a truly random number. Finally, Eve_{link} guesses r and outputs r'.

$$Advantage_{Eve_{link}} = |Pr[r' \longleftarrow Eve_{link}, r = r'] - 1/2|$$

In the proposed system, the advantage of any probabilistic polynomial-time of adversary Eve_{link} is negligible. Therefore, the proposed system is unlinkable, i.e., it is impossible to find a relationship among the output of T through the cooperation of DBs.

In the case of the DB, and without knowing the UID of the tag, it is impossible to compute MID_T stored in other DBs. This is because each DB is stored with different MID_T, and all MID_Ts are generated differently.

When the reader attempts to perform tracking by collecting transmitted data from the tag, it is impossible to ensure that the transmitted data from the same tag will change for each query.

Theorem 2. *Assuming functions H is random oracles, the proposed system is* ***unlinkable****.*

The proposed system is proven to ensure users' privacy, compared with the existing methods since the response from the tag is one-time based information. In addition, the UIDs stored in the IMC are encrypted with K, and K is consigned to each of the IMC administrator in the secret sharing scheme for secure-keeping.

To track the UID from the tag in *IMC*, the original secret may be recovered through the cooperation of $t(t \leq n)$ or more of the consigned institutions among "n" consigned institutions. Accordingly, in case there is a need to trace the UID in the tag, the UID can be acquired by recovering the key used in encryption only through the cooperation of "t" administrators or more. As a result, the privacy of the users can be protected better.

6   Conclusion

The RFID system have become popular for automated identification and supply-chain applications. Furthermore, the RFID system can be applicable to location based systems. In location providing services such as a search for missing children, and emergency call services, the availability of the user location information is critical in providing specific location information. However, location is a sensitive piece of information, and the user usually desires protect their location information secret. Therefore, when necessary, only authorized entities should have access to location information.

In this paper, a secure and fair RFID system is proposed. The proposed system can authenticate the tag without exposing its UID to the reader (also the back-end database), and therefore protect users' privacy. However, if necessary, only authorized administrators can identify the UID of the tag.

In conclusion, the following two points are stated; (1) The tag generates a random response with a random value received from the reader on every session. Therefore, the proposed system secures against an adversary attempting a replay attack, tracking, and so on. (2) The UID of the tag can be traced, when necessary, through the cooperation of authorized administrators. Therefore, the proposed RFID system can be satisfied with *fairness.*

References

1. C. Cachin. On-Line Secret Sharing. Cryptography and Coding: 5th IMA Conference, LNCS 1025, pp. 190-198, Springer-Verlag, 1995.
2. L. Chen, D. Gollmann, C. J. Mitchell and P. Wild. Secret sharing with Reusable Polynomial. 2nd Australasian Conference on Information Security and Privacy, ACISP 97, LNCS 1270, pp. 183-193, Springer-Verlag, 1997.
3. D. Engels. The Reader Collision Problem. Technical Report. MIT-AUTOID-WH-007, MIT Auto ID Center, 2001. Available from http://www.autoidcenter.org.
4. D. M. Ewatt and M. Hayes. Gillette razors get new edge: RFID tags. Information Week, 13 January 2003. Available from http://www.informationweek.com.
5. P. Fedlman. A Practical scheme for Non-interactive Verifiable secret sharing. 28th Annual Symposium on the Foundation of Computer Science, pp. 427-437, 1987.
6. K. Finkenzeller. RFID Handbook, John Wiley and Sons. 1999.
7. G. Avoine and P. Oechslin. A Scalable and Provably Secure Hash-Based RFID Protocol. 2nd IEEE International Workshop on Pervqsive Computing and Communications Security, PerSec 2005, pp. 110-114, IEEE, 2005.
8. G. Avoine and P. Oechslin. RFID Traceability: A Multilayer Problem. Financial Cryptography, FC'05, LNCS 3570, pp. 125-140, Springer-Verlag, 2005.

9. A. Juels and R. Pappu. Squealing Euros: Privacy protection in RFID-enabled ban-knotes. Financial Cryptography, FC'03, LNCS 2742, pp. 103-121, Springer-Verlag, 2003.

10. A. Juels, R. L. Rivest and M. Szydlo. The Blocker Tag : Selective Blocking of RFID Tags for Consumer Privacy. 10th ACM Conference on Computer and Communications Security, CCS 2003, pp. 103-111, 2003.

11. S. J. Kim, S. J. Park and D. H. Won. Proxy Signatures, Revisited. International Conference on Information and Communications Security, ICISC'97, LNCS 1334, pp. 223-232, Springer-Verlag, 1997.

12. H. Knospe and H. Pobl. RFID Security. Information Security Technical Report, vol. 9, issue 4, pp. 39-50, Elsevier, 2004.

13. MIT Auto-ID Center (EPCglobal), http://www.epcglobalinc.org.

14. M. Ohkubo, K. Suzuki, and S. Kinoshita. A Cryptographic Approach to "Privacy-Friendly" tag. RFID Privacy Workshop, Nov 2003. http://www.rfidprivacy.org/

15. T. P. Pedersen. A Threshold cryptosystem without a trusted party. Advances in Cryptology-EUROCRYPT'91: Workshop on the Theory and Application of Cryptographic Techniques, LNCS 547, pp. 522-526, Springer-verlag, 1991.

16. K. W. Rhee, J. Kwak, S. J. Kim, and D. H. Won. Challenge-Response Based RFID Authentication Protocols for Distributed Database Environment. Second International Conference on Security in Pervasive Computing, SPC 2005, LNCS 3450, pp. 70-84, Springer-Verlag, 2005.

17. S. E. Sarma. Towards the five-cent tag. Technical Report MIT-AUTOID-WH-006, MIT Auto ID Center, 2001. Available from http://www.autoidcenter.org.

18. S. E. Sarma, S. A. Weis, and D. W. Engels. RFID systems, security and privacy implications. Technical Report MIT-AUTOID-WH-014, AutoID Center, MIT, 2002.

19. S. E. Sarma, S. A. Weis, and D. W. Engels. Radio-frequency identification systems. Workshop on Cryptographic Hardware and Embedded Systems, CHES'02, LNCS 2523, pp. 454-469, Springer-Verlag, 2002.

20. S. E. Sarma, S. A. Weis, and D. W. Engels. Radio-frequency-identification security risks and challenges. CryptoBytes, 6(1), 2003.

21. T. Scharfeld. An Analysis of the Fundamental Constraints on Low Cost Passive Radio-Frequency Identification System Design. MS Thesis, Department of Mechanical Engineering, Massachusetts Institute of Technology, Cambridge, MA 02139, 2001.

22. Security technology: Where's the smart money? The Economist, pages 69-70, 9 February 2002.

23. A. Shamir, How to share a secret. Communication of the ACM, vol. 21, pp. 120-126, 1979.

24. M. Tompa and H. Woll. How to share a secret with cheater. Journal of Cryptology, vol. 1, pp. 133-138, 1988.

25. I. Vajda and L. Buttyan. Lightweight Authentication Protocols for Low-Cost RFID Tags. 2nd Workshop on Security in Ubiquitous Computing, Ubicomp 2003, 2003.

26. S. A. Weis. Radio-frequency identification security and privacy. Master's thesis, M.I.T. May 2003.

27. S. A. Weis, S. Sarma, R. Rivest, and D. Engels. Security and privacy aspects of low-cost radio frequency identification systems. First International Conference on Security in Pervasive Computing, SPC 2004, LNCS 2802, pp. 201-212, Springer-Verlag, 2004.

Scalable and Flexible Privacy Protection Scheme for RFID Systems

Sang-Soo Yeo and Sung Kwon Kim

Chung-Ang University, Seoul, Korea
ssyeo@alg.cse.cau.ac.kr, skkim@cau.ac.kr

Abstract. Radio Frequency Identification (RFID) system has been studied so much and it may be applicable to various fields. RFID system, however, still has consumer privacy problems under the limitation of low-cost tag implementation. We propose an efficient privacy protection scheme using two hash functions in the tag. We show that our scheme satisfies not only privacy and location history protection of consumers, but also scalability and flexibility of back-end servers. Additionally, we present a practical example to compare performance of several schemes.

1 Introduction

Recently, in the areas of manufacturing, logistics, sales and consumer services, radio frequency identification(RFID) system has been adopted rapidly. This trend may be caused of cost-down of manufacturing RFID tag, which is about $5 \sim 10$ cents through large purchase orders. And accelerated adoption of RFID systems of multinational enterprises and of national agencies plays an important role in widely use of RFID systems. But low cost RFID tag has important security holes such as information leakage, location tracking, location history disclosure and counterfeiting [1][2][3]. Particularly, because low cost RFID tag gives an identical answer to any readers, privacy of consumer bearing item, in which RFID tag embedded, may be not guaranteed and location of him may be traceable. In low cost RFID tag, it is infeasible to embed high level cryptographic primitives such that public-key cryptography or qualified PRNG(pseudo-random number generator) [3]. Only hash functions or block ciphers seem to be implemented in low cost RFID tag [4].

In this paper, we propose a new privacy protection scheme that guarantees privacy protection and location history protection for consumers and that improves performance of back-end server. Our scheme is similar to the scheme of Ohkubo et al. [1] in the view of self-refreshment of the tag. And our scheme provides indistinguishability and forward security and guarantees reasonable complexity of back-end server computation. In section 2, we mention some issues related to privacy and security in RFID systems. In section 3, we summarize several schemes for privacy protection. In section 4, we present our basic scheme for privacy protection. In section 5, we introduce our main scheme that enhances tag's security and server's performance. And then we describe the pre-computation

R. Molva, G. Tsudik, and D. Westhoff (Eds.): ESAS 2005, LNCS 3813, pp. 153–163, 2005.
© Springer-Verlag Berlin Heidelberg 2005

approach for the main scheme. In section 6, with a practical example, our scheme is evaluated and compared to other schemes' results. In section 7, we conclude with mentioning the significance of our scheme and results.

2   Privacy and Security Issues in RFID System

2.1   RFID System

Normally, RFID system is consist of four main components [1][5]; tag, reader, data processing subsystem, and back-end server.

The RFID *tag*, or *transponder*, is embedded in the item(object) to be identified. The tag comprises an IC chip and an antenna module. The passive tag, which has not own battery for reducing the manufacturing cost, receives reader's query through radio signal and sends its answer to reader using harvested energy from the electromagnetic field of the reader's radio signal. The RFID *reader* communicates several tags at the same time and identifies their ID through data processing subsystem and/or back-end server system. The *data processing subsystem* is attached to the reader and retrieves appropriate information from its own database or external database server according to the data obtained by the reader. Generally, the data processing subsystem is considered as a part of reader. The *back-end server* has a database and manages various types of information related to each tag. The answer of the tag is transmitted securely to the back-end server through authenticated reader and it is used to identify the tag. The back-end server must be trusted and must have the capability to process every query from a lot of readers concurrently.

2.2   Privacy Problems and Security Requirements

RFID Privacy problem has two main issues. Therefore two security requirements are needed for solving the problem [1]. These are described below.

Privacy Problems. *Information Leakage* - RFID tag gives data to any readers. The privacy of the consumer, who has tag-embedded items, should not be guaranteed. If anyone have a RFID reader, he can acquire tag data easily and guess the owner's personality. *Traceability* - If the tag always answers identical data such as unique ID, its owner's location can be traced continuously.

Security Requirements. *Indistinguishability* - This means that the output of the tag must be indistinguishable from truly random values and the output should be unlinkable to its own ID [1]. If the output of the tag is always indistinguishable, it guarantees the complete prevention of information leakage and the partial prevention of traceability. *Forward Security* - This means that even if the adversary, who has a set of readings between tags and readers, acquires the secret data stored in the tag, he cannot find the relation to past events in which the tag was involved [1]. Because the tag can be tampered physically, this requirement is important in RFID privacy problem. This requirement guarantees the complete untraceability.

2.3   Requirements of Back-End Server

Efficiency and Scalability. Some privacy protection schemes [1][6] require that back-end server performs a large amount of computation for each identification request. Some other schemes [9][10] require that the reader frequently writes new data on the tag. In these case, computations of the reader or the back-end server must be efficient and scalable.

Flexibility. The database of back-end server must be flexible. In some approaches [11], back-end server must rebuild its database whenever a tag is registered or unregistered. It takes very long time to rebuild the database.

3   Related Work

Recently, many researches related to RFID privacy problem are published. Sarma et al. announced privacy issues of RFID tag and introduced *Self Destruct* command [5]. Weis et al. suggested hash lock scheme, randomized hash lock scheme, and silent tree walking scheme [3]. Juels et al. introduced *Blocker tag* [7] and suggested privacy protection scheme for RFID-enabled banknotes [2]. Golle et al. suggested re-encryption scheme that reader writes frequently new secret on the tag using ElGamal public-key cryptosystem [9] and Saito et al. proposed modified scheme of [9]. Henrici et al. suggested hash based scheme using transaction counter [8].

Unfortunately, all above schemes are insecure. For the detailed descriptions of security holes and possible attacks for them, refer to [1] and [11]. Particularly, the scheme of Saito et al. [10] seems to be infeasible for a low cost tag because the tag must perform ElGamal decryption by itself.

However, the basic scheme Ohkubo et al. is secure [1]. In this scheme, the tag refreshes its secret data by itself using two one-way hash functions, while the reader refreshes secret data of the tag in most of above schemes. Nevertheless, this scheme is impractical and not scalable because it has a very high computational complexity of back-end server to resolve the ID of the tag. (The scheme of Ohkubo et al. has a complexity of $O(mn)$, where m is the number of tags and n is the maximum length of hash-chains.) Though Ohkubo et al. suggested some enhancing techniques in [6], which decreased security of the original scheme partially. Avoine et al. proposed the pre-computation approach supporting the basic scheme of Ohkubo et al. and makes it practical [11]. In the approach of Avoine et al., however, the pre-computation table must be rebuilt whenever a new tag is registered on back-end server or a useless tag is removed from back-end server. It takes very long time for the pre-computation.

4   Basic Scheme

RFID System Construction.

H : $\{0,1\}^* \longrightarrow \{0,1\}^\ell$, the one-way hash function algorithm with outputs of length ℓ.

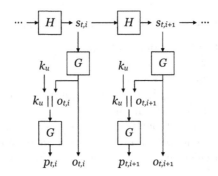

Fig. 1. The refreshment and the output of the tag in the basic scheme

G : $\{0,1\}^* \longrightarrow \{0,1\}^\ell$, the one-way hash function algorithm with outputs of
   length ℓ. G has the different distribution from that of H.
m : the number of tags.
n : the maximum length of the hash chain for each tag.
g : the number of tag groups.
T_t : the tag has the hash seed, the group index, and two types of hash functions.
R : the reader communicates with B through a secure channel.
B : the back-end server manages a database of $(id_t, s_{t,1})$, in which $s_{t,1}$ is created
   randomly for each tag. Additionally, B manages a database of tag groups
   and group index value k_u.
id_t : the ID of tag t, where $t = 1, 2, \cdots, m$.
$s_{t,1}$: the hash seed of tag t.
$s_{t,i}$: the i-th hash chain value of tag t, where $i = 1, 2, \cdots, n$.
k_u : the index value of tag group u, where $u = 1, 2, \cdots, g$.

4.1   Scheme Outline

Main concerns of the basic scheme are in satisfying the forward security using
two hash functions in the tag and in reducing the computational complexity of
back-end server using the group index in the tag and back-end server. Whenever
it is queried by a reader, the tag computes output data $p_{t,i} || o_{t,i}$ for the reader
using hash function G and the tag changes its secret data $s_{t,i}$ using hash function
H (fig.1). As mentioned above, the scheme using two hash functions is originally
proposed by Ohkubo et al. [1]. We have modified their scheme and added the
step of hashing the group index concatenated to $o_{t,i}$ together. The group index
is a fixed value of each group and all of m tags are evenly divided into g groups,
and so the number of tags in each group is m/g. Back-end server finds k_u by
checking $p_{t,i} = G(k_u || o_{t,i})$ for all $1 \le u \le g$ and then finds $(id_t, s_{t,1})$ by checking
$o_{t,i} = G(H^{i-1}(s_{t,1}))$ for all $1 \le i \le n$ and all $t \in$ group u.

Detailed Protocol.

1. Reader, R
   (a) sends request to the tag.

2. Tag, T_t
    (a) computes $o_{t,i} = G(s_{t,i})$ and $p_{t,i} = G(k_u \,||\, o_{t,i})$.
    (b) sends answer $p_{t,i} \,||\, o_{t,i}$ to the reader.
    (c) changes secret data $s_{t,i+1} = H(s_{t,i})$.

3. Reader, R
    (a) receives $p_{t,i} \,||\, o_{t,i}$ from T_t.
    (b) sends $p_{t,i} \,||\, o_{t,i}$ to B through a secure channel.

4. Back-end Server, B
    (a) receives $p_{t,i} \,||\, o_{t,i}$ from the authenticated R.
    (b) finds k_u by checking $p_{t,i} = G(k_u || o_{t,i})$ for all $1 \le u \le g$.
    (c) finds $(id_t, s_{t,1})$ by checking $o_{t,i} = G(H^{i-1}(s_{t,1}))$ for all $1 \le i \le n$ and all $t \in$ group u.
    (d) sends id_t to the authenticated R through a secure channel.

5. Reader, R
    (a) receives id_t of T_t from B.

4.2 Security and Performance Analysis

Security. Firstly, the output of hash function G is indistinguishable, hence this scheme satisfies indistinguishability. Even if the tag is tampered and $s_{t,i}$ is revealed by adversary, he will not know $s_{t,j}$, for $j < i$, because of one-way property of hash function H. However, if k_u is revealed to him, forward security will not be guaranteed. He can obtain location history of the tag by computing with k_u and prior outputs of the tag. This is adjusted in the our main scheme.

Performance. The complexity of hashing in the step of finding group u to which the tag belongs is $O(g)$ and the complexity of hashing in the step of finding $s_{t,1}$ in group u is $O\left(n\frac{m}{g}\right)$. Therefore the total complexity in terms of hashing, T_{basic}, is $O\left(g + \frac{mn}{g}\right)$. If $g = \sqrt{mn}$, T_{basic} is minimized and $T_{basic} = (\sqrt{mn})$. The scheme of Ohkubo et al. has the complexity of $O(mn)$ [1].

5 Main Scheme

RFID System Construction. It is identical to the basic scheme except below.

l : the maximum length of the hash chain for each group. We assumes $l = n$.

B : the back-end server manages a database of $(id_t, s_{t,1})$, in which $s_{t,1}$ is created randomly for each tag. And B manages a database of tag groups and each group seed, $k_{u,1}$.

$k_{u,1}$: the seed of tag group u, where $u = 1, 2, \cdots, g$.

$k_{u,i}$: the i-th hash chain value of tag group u, where $i = 1, 2, \cdots, n$.

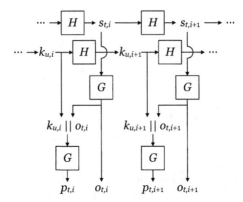

Fig. 2. The refreshment and the output of the tag in the main scheme

5.1   Scheme Outline

In the main scheme, we replace group index k_u in the basic scheme by another hash chain(fig.2). This modification gives a stronger forward security to our scheme. The tag has only two hash functions H and G, therefore the tag's hardware is equivalent to the scheme of Ohkubo et al. in the view of the number of hash modules. Whenever it is queried by a reader, the tag computes output data $p_{t,i}||o_{t,i}$ and the tag changes its secret data $s_{t,i}$ and its group hash value $k_{u,i}$. $p_{t,i}$ is computed from $o_{t,i}$ and $k_{u,i}$. Back-end server finds $k_{u,1}$ by checking $p_{t,i} = G(H^{i-1}(k_{u,1})||o_{t,i})$ for all $1 \le u \le g$ and all $1 \le i \le n$. And then B finds $(id_t, s_{t,1})$ by checking $o_{t,i} = G(H^{i-1}(s_{t,1}))$ for all $t \in$ group u.

Detailed Protocol.

1. Reader, R
   (a) sends request to the tag.

2. Tag, T_t
   (a) computes $o_{t,i} = G(s_{t,i})$ and $p_{t,i} = G(k_{u,i} || o_{t,i})$.
   (b) sends answer $p_{t,i} || o_{t,i}$.
   (c) changes secret data $s_{t,i+1} = H(s_{t,i})$ and changes group hash value $k_{u,i+1} = H(k_{u,i})$.

3. Reader, R
   (a) receives $p_{t,i} || o_{t,i}$ from T_t.
   (b) sends $p_{t,i} || o_{t,i}$ to B through a secure channel.

4. Back-end Server, B
   (a) receives $p_{t,i} || o_{t,i}$ from the authenticated R.
   (b) finds $k_{u,1}$ and i by checking $p_{t,i} = G(H^{i^*-1}(k_{u,1})||o_{t,i})$ for all $1 \le u \le g$ and all $1 \le i^* \le n$.
   (c) finds $(id_t, s_{t,1})$ by checking $o_{t,i} = G(H^{i^*-1}(s_{t,1}))$ for all $t \in$ group u.
   (d) sends id_t to the authenticated R through a secure channel.

5. Reader, R
   (a) receives id_t of T_t from B.

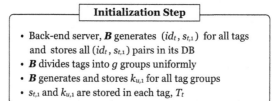

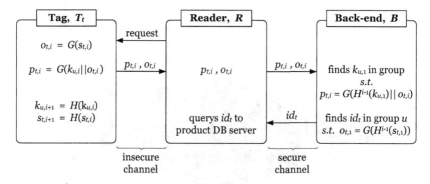

Fig. 3. The overview of our main scheme

5.2   Security and Performance Analysis

Security. As mentioned in the basic scheme, the output of hash function G is indistinguishable, hence this scheme satisfies indistinguishability. Even if the tag is tampered and $s_{t,i}$ is revealed by adversary, he will not know $s_{t,j}$, for $j < i$, because of one-way property of hash function H. And the group hash value in the tag is also refreshed by hash function H whenever the tag sends answer. Even if the tag is tampered and $k_{u,i}$ is revealed by the adversary, he will not know $k_{u,j}$, for $j < i$. Therefore forward security is guaranteed.

Additionally, we consider the case that the adversary who tampered one tag, T_a and knows $k_{u,i}$ would try to trace another tag, T_b. Because he does not know T_a's i, T_b's group hash seed, and T_b's transaction counter i', he cannot distinguish T_b from the others. Even in the case that T_b and T_a are in the same group, only if $i' < i$, he cannot trace it. However, if and only if T_b and T_a are in the same group and $i' \geq i$, he can distinguish T_b from the others by computing $p_{t_b,i'} = G(k_{u,i'}||o_{t_b,i'})$ for all $i \leq i' \leq n$. Nevertheless, if the number of groups is large enough, such as $\frac{m}{4} \sim \frac{m}{2}$, the number of tags in the same group will be very small and the security risk will be reduced considerably.

Performance. The complexity of hashing in the step of finding group u is $O(ng)$ and the complexity of hashing in the step of finding $s_{t,1}$ in group u is $O\left(n\frac{m}{g}\right)$. Therefore the total complexity in terms of hashing, T_{main}, is $O\left(ng + n\frac{m}{g}\right)$. If $g = \sqrt{m}$, T_{main} is minimized and $T_{main} = O(n\sqrt{m})$, but it is not enough to ensure the security . Practically, it is recommended that g should be large enough for a strong security.

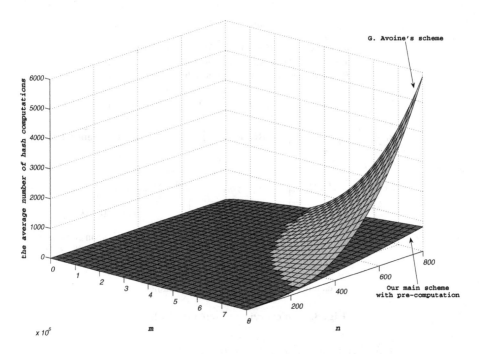

Fig. 4. The average number of hash computations, where $M = 2^{30}$ (1GB Memory)

The flexibility of back-end server is good because it is easy to add a new tag into back-end server's DB and to remove an useless tag.

Performance Enhancement with Pre-computation. We can build pre-computation table for the step of finding group u using the pre-computation technique of Avoine et al. [11]. This technique has no effect on flexibility of our scheme, because the number of groups is not changed when a new tag is registered or an useless tag is removed. Using the pre-computation technique of Avoine et al. [11] with M bytes memory, adequate g value can be computed as below (proof is omitted):

$$g = \left(\frac{mM^3\mu^2}{3^4 2^9 n^2 \gamma} \right)^{\frac{1}{4}} \text{, where we assume } \gamma \text{ and } \mu \text{ small factors. Refer them to [11]}$$

So we can compute $T_{main\_with\_pre-computation}$ straightforwardly:

$$T_{main\_with\_pre-computation} = \left(\frac{2^{17} m^3 n^6 \gamma}{M^3 \mu^2} \right)^{\frac{1}{4}} = O\left(\left(\frac{m^3 n^6}{M^3} \right)^{\frac{1}{4}} \right)$$

The complexity of the scheme of Ohkubo et al. is $O(mn)$ [1] and the complexity of the scheme of Avoine et al. is $\frac{3^3 2^9 n^3 m^3 \gamma}{M^3 \mu^2} = O\left(\frac{m^3 n^3}{M^3} \right)$ [11]. Therefore our main scheme with pre-computation has the lower complexity than Ohkubo et al., where $m > \frac{n^2}{M^3}$. And it has the lower complexity than Avoine et al., where $m > \frac{M}{\sqrt[3]{n}}$. Fig.4 shows comparison of our main scheme with pre-computation and the scheme of Avoine et al. in the view of the average number of hash computations. A practical example is introduced in the next section.

6 Practical Example

We compare the scheme of Ohkubo et al. , the scheme of Avoine et al. and our scheme in the view of hash computation complexity of back-end server.

- Example setting:

   $m = 2^{20}$: the number of tags

   $n = 2^{10}$: the maximum length of each hash chain

   g: the number of group

   $\text{hash}_{speed} = 2^{24} times/sec$: the number of times of hash function per seconds

   M: the amount of memory for pre-computation table (bytes)

   $\mu = 2$: refer to [11]

   $\gamma = 8$: refer to [11]

- the scheme of Ohkubo et al.

$$\text{complexity} = nm = 2^{30}$$
$$\text{identification time} = 2^6 sec. = 64sec.$$

- the scheme of Avoine et al. ($M = 2^{30}$ bytes)

$$\text{complexity of pre-computation} \approx \frac{n^2m}{2} = 2^{39}$$
$$\text{pre-computation time} \approx 2^{15} sec. \approx 10hours$$
$$\text{complexity of identification} \approx \left(\frac{3^3 2^9 n^3 m^3 \gamma}{M^3 \mu^2} \right) = 2^8 \cdot 3^3$$
$$\text{identification time} = 3^3 \cdot 2^{-14} sec. \approx 0.00165sec.$$

- our basic scheme ($g = \sqrt{nm}$)

$$\text{complexity} = \sqrt{nm} = 2^{15}$$
$$\text{identification time} = 2^{-9} sec. \approx 0.00195sec.$$

- our main scheme without pre-computation ($g = \sqrt{m}$)

$$\text{complexity} = n \cdot \sqrt{m} = 2^{20}$$
$$\text{identification time} = 2^{-4} sec. \approx 0.0625sec.$$

− our main scheme with pre-computation ($M = 2^{27}bytes$)

$$\text{the optimum of } g = \left(\frac{mM^3\mu^2}{3^4 2^9 n^2 \gamma}\right)^{\frac{1}{4}}$$

$$= 2^{17} \cdot 2^{\frac{3}{4}} \cdot 3^{-1} \approx 2^{17}$$

$$\text{complexity of pre-computation} \approx \frac{gn^2}{2} \approx 2^{36}$$

$$\text{pre-computation time} \approx 2^{12} sec. \approx 69min.$$

$$\text{complexity of identification} \approx \left(\frac{2^{17} n^6 m^3 \gamma}{M^3 \mu^2}\right)^{\frac{1}{4}} = 2^{13}$$

$$\text{identification time} = 2^{-11} sec. \approx 0.00049 sec.$$

7   Conclusion

The scheme of Ohkubo et al. is secure, but not scalable. We have described server-side computational advantages of grouping the tags in our basic scheme. And we have proposed the scalable and flexible privacy protection scheme that satisfies indistinguishability and forward security. We have shown that our main scheme has a enhanced identification time in back-end server without pre-computation compared to Ohkubo et al.'s scheme. Applying the pre-computation technique of Avoine et al. to the step of finding the group in our main scheme, we have shown that our main scheme can be more efficient and more scalable.

Acknowledgement

This work was supported by grant No. R01-2005-000-10568-0 from the Basic Research Program of the Korea Science & Engineering Foundation.

References

1. M. Ohkubo, K. Suzuki, and S. Kinoshita, "Cryptographic Approach to "Privacy-Friendly" Tags". *RFID Privacy Workshop*, MIT, November 2003.
2. A. Juels and R. Pappu, "Squealing Euros: Privacy Protection in RFID-Enabled Banknotes", *Financial Cryptography '03*, 2003.
3. S. Weis, S. Sarma, R. Rivest, and D. Engels, "Security and Privacy Aspects of Low-cost Radio Frequency Identification Systems", In *Proceedings of the 1st International Conference on Security in Pervasive Computing*, 2003.
4. M. Feldhofer, "A Proposal for Authentication Protocol in A Security Layer for RFID Smart Tags", *IEEE MELECON 2004*, vol. 2, pp. 759-762, May 2004.
5. S. Sarma, S. Weis, and D. Engels, "RFID Systems and Security and Privacy Implications", *CHES 2002*, vol. 2523 of LNCS, pp. 454-469, August 2002.
6. M. Ohkubo, K. Suzuki and S. Kinoshita, "Efficient Hash-Chain Based RFID Privacy Protection Scheme", *Ubicomp 2004*, 2004.

7. A. Juels, R. Rivest, and M. Szydlo, "The Blocker Tag : Selective Blocking of RFID Tags for Consumer Privacy", *ACM CCS 2003*, pp. 27-30, October 2003.
8. D. Henrici, and P. Müller, "Hash-based Enhancement of Location Privacy for Radio-Frequency Identification Devices using Varying Identifiers", *IEEE PerSec '04* at *IEEE PerCom*, March 2004.
9. P. Golle, M. Jakobsson, A. Juels, and P. Syverson, "Universal Re-Encryption for Mixnets", *CT-RSA '04*, 2004.
10. J. Saito, J-C. Ryou and K. Sakurai, "Enhancing Privacy of Universal Re-Encryption Scheme for RFID Tags", *EUC '04*, pp. 879-890, August 2004.
11. G. Avoine and P. Oechslin. "A Scalable and Provably Secure Hash-Based RFID Protocol", *IEEE PerSec 2005*, Kauai island, Hawaii, USA, March 8th, 2005.

RFID Authentication Protocol with Strong Resistance Against Traceability and Denial of Service Attacks

Jeonil Kang and DaeHun Nyang

Information Security Research Laboratory,
INHA University*, Korea
dreamx@seclab.inha.ac.kr, nyang@inha.ac.kr
http://seclab.inha.ac.kr

Abstract. Even if there are many authentication protocols for RFID system, only a few protocols support location privacy. Because of tag's hardware limitation, these protocols suffer from many security threats, especially from the DoS (Denial of Service) attacks. In this paper, we discuss location privacy problem and show vulnerabilities of RFID authentication protocols. And then, we will suggest a strong authentication protocol against location tracing, spoofing attack, and DoS attack.

1 Introduction

A radio-frequency identification (RFID) system has been widely deployed mainly in supply chain management. Compared to using optical bar-code, RFID system has many benefits: quick reading, long recognition distance, obstacle-free, strength against the contamination, etc. Owing to these properties, a lot of tags can be read simultaneously during a few seconds. Also, RFID system can be effectively used in some applications such like animal tagging, high-way tolling, theft protecting, etc, whereas bar-code cannot handle these sides. RFID system will replace bar-code system very quickly.

Unfortunately, RFID system also has too many security risks specially when a high level of security is required. Generally, RFID tag has very low resources: low computing power and small memory size. Thus, it is very hard to apply existing security technologies that assumes very high computing power and large memory size to RFID tag. So, a lot of researches have been considered about security techniques for low-cost RFID systems [1-9]. We can classify security problems of RFID systems into two categories: information leakage and traceability.

A passive attacker might be able to overhear the information between a reader and some tags because the medium is the air in the RFID system. An active attacker may be able to send some bogus data that fakes the reader or tags to extract information from them. To prevent these attacks, many protocols have

---

* This work was supported by INHA UNIVERSITY Research Grant.

R. Molva, G. Tsudik, and D. Westhoff (Eds.): ESAS 2005, LNCS 3813, pp. 164–175, 2005.
© Springer-Verlag Berlin Heidelberg 2005

been proposed: blocker tag [6], RFID system using AES [9], minimalist XOR one-time cryptography [4], etc. But still the blocker tag is a very simple and strong method to protect information leakage.

On the other hand, tag's location information as well as tag's own data must be protected. Also, there are some solutions for this problem; hash-based ID variation protocol [1], hash chaining [2], universal re-encryption [7], etc. According to G. Avoine, however, it is hard to protect the threat because each RFID protocol layer (application, communication, and physical layer) has technical and actual defects [10].

In this paper, we'll discuss the location privacy of RFID system in section 2. Vulnerability of existing models of authentication protocols to location tracing and the DoS attack is shown in section 3. In section 4, we'll propose a strong authentication protocol for RFID system to solve the different early mentioned problems. Finally, we will summarize our results in section 5.

2   Location Privacy

2.1   Traceability at the Application Layer

Generally, RFID communication protocol consists of three layers; application, communication and physical layer. In the application layer, the information is handled by the user. Other RFID application programs use this information to identify objects. To make this possible, generally the information is made of product number and serial number. Also it might have a secret key or password for administrative purpose. The communication layer defines how to avoid collision might occur by the reader or the tags. This protocol is called "Collision Avoidance Protocol" or "Anti-Collision Algorithm." Collision avoidance protocol can be classified into probabilistic and deterministic methods according to the predictability of singulation time. The physical layer defines how to transmit data physically. (e.g. air interface - frequency, modulation, data encoding, synchronization, etc.)

Unfortunately, location privacy problem might occur in each layer. If an attacker can overhear all messages between the reader and tags which contents contain information that is never changed, the attacker can trace tag's movement using the information. Also, the attacker can use a type of collision avoidance protocol or tag identifier which can be used in collision avoidance protocol. Therefore, it is hard to prevent the attacker from tracing.

The one thing that we must consider to protect location privacy is accuracy (or correctness) of tracing. Accuracy of tracing using defects in communication and physical layer is lower than that in application layer. That's because there are many tags which use the same collision avoidance protocol and have same physical features in the real world. However, accuracy of tracing with messages in application layer is very high, because the databases server should distinguish each tag clearly. Therefore, it is very important to protect location privacy in the application layer. In this paper, we will propose a method to solve the location privacy problem in application layer.

2.2   Threat Model

Before describing our authentication protocol, a threat model is defined to analyze vulnerabilities of existing authentication model. By defining the threat model, we can restrict the ability of the attacker reasonably in RFID system.

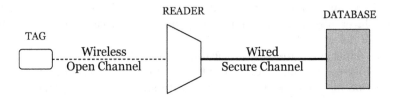

Fig. 1. RFID system threat model. An attacker cannot intercept in wireless channel because of a property of the air. Also, we assume that wired channel is secure in any case.

Like another threat model, we assume that an attacker can't intercept or modify messages. In the wireless open channel between tag and reader, the attacker can overhear all messages or insert fake messages, but he can't intercept or modify messages because the medium of transmission is the air itself. In wired secure channel between reader and database, an attacker can't eavesdrop, intercept, insert, or modify. In RFID system, we assume that reader and database server have pre-authenticated each other.

In the threat model, the attacker tries to trace location of tags and also tries to collapse system. The system collapse is caused by DoS(Denial of the Service), spoofing, or asynchronisation attack against specific, unspecific tags or against the database.

3   Vulnerabilities of Authentication Protocols

In this section, we will describe some methods to attack RFID system using structural problem of existing authentication protocols.

3.1   Un-terminated Session Attack

We cannot be sure that an attacker may always terminate the authentication session correctly. The authentication protocols - which send only hashed identifier (or hashed identifier with the nonce that was received from a reader) to a reader - are more prone to be traceable. Since the tag returns same hashed identifier in every session for attacker's queries, attacker can easily find tag's location. Unfortunately, the tag can't change its identifier to prevent this attack because the identifier always must be synchronized with that in the database. This asynchronisation (between database and tag) means that we cannot use the tag anymore.

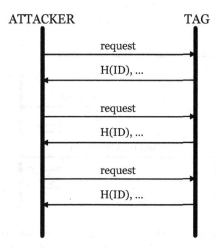

Fig. 2. Location tracing using un-terminated session. An attacker can trace target tag by sending request message and closing session.

3.2 Preemptive Locking

Consider that a tag receives a new request message, while it is already in the authentication session. The tag should choose its action: to ignore the request or to start immediately a new authentication session for the new request. The former gives a chance for the attacker to lock a tag preemptively. The attacker can make a target tag keep silence just by sending request message to the tag before it starts a legal authentication session. Because of this preemptive locking, the manufacturer should provide some mechanism with the tags to prevent this attack, or choose the latter one as the reaction of the tag.

Using timer, it looks rather easy to prevent preemptive locking. A tag has a timer and it starts the timer when session starts. When the timer is expired, the tag closes the session instantly. However, during the time that attacker possesses the session, tag cannot serve any request and thus, the attack might cause a severe degradation of the overall performance.

So, it seems to be better to start new authentication session for the newly arrived request message. But even in this case, tag is still weak against the attack described in section 3.3.

3.3 Stealth Bombing

If a tag is implemented to start new authentication session for the newly arrived request message, then it will suffer from so called, 'stealth bombing'. Stealth bombing is a kind of the DoS attacks. We assumed that the attacker can generate and insert fake messages, in section 2. What will happen if fake messages are inserted by the attacker during authentication session? If the tag which changes its response at every session using a random nonce opens a new session for a fake request message, the ongoing legal authentication session will fail by this illegal authentication trial.

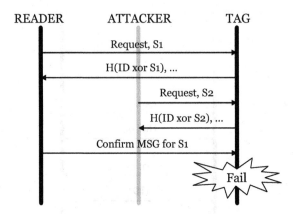

Fig. 3. Stealth bombing attack. An attacker can attempt DoS attack by sending request message to the tag in session. The tag has to decide its action against the request message.

Also, we can think another attacking method. Instead of sending a fake request message, the attacker might send an invalid confirm message for a legal session request. Some type of tags might abort the authentication session after receiving this confirm message.

Though this stealth bombing attack seems to be very hard to try because the communication layer will discard if the frame does not have a valid identifier defined by the communication layer, it is possible to send a fake request message that includes the valid identifier since it is disclosed during the normal communication between reader and tag.

3.4   DoS Attack Against the Database

Some authentication protocols use status-based flag to prevent problems of section 3.1. If the previous session terminated unsuccessfully, the tag responds using the message created with spare key and hint for the real identifier. These protocols must use large computational resource for recovering correct status and identifier. For another example, Ohkubo's protocol [2] must calculate a lot of hash function to search tag's identifier because the database server must compute $s_i = h(h(\cdots h(s_1)\cdots))$ from s_1. Since RFID tag has only small computing power, their approach to move tag's computation to database server is reasonable.

Because of this computational inefficiency, the attacker can try the DoS attack with very small effort against these protocols. It is enough to send some garbage messages toward the database server. Then, the database server will start to search identifier until finding it or retrieving all records. The attacker doesn't need to check whether the identifier exists or not when he makes fake messages.

In conclusion, a strong authentication protocol against DoS attacks should not search a large space for finding the hidden identifier. It seems to be incompatible to the solution of the problems in section 3.1.

4   Proposed Protocol

In this section, we will propose an authentication protocol which has a tolerance against DoS attack, provides location privacy, and strengthens other weakness.

4.1   Tag and Database Structure

A tag using this protocol has fields shown in below.

- SFlag (session flag): indicates whether tag is under an authentication session or not. When a session starts, it is set to true, and when a session ends either normally or abnormally, it is set to false. SFlag is initialized to false immediately after a tag comes to be active.
- ID (identifier): used to distinguish a tag from another tags during authentication session. $\{0, 1\}^n$
- CWD (confirm word): used to confirm the ID of tag by the database. $\{0, 1\}^n$
- R1 (Random nonce 1): changed at each session. If SFlag is true, it is replaced by saved R1. Or it is generated by the tag newly. $\{0, 1\}^n$
- C (Counter): increased or randomly changed at each session. If SFlag is true, it is replaced by saved C. Or it is changed by the tag. $\{0, 1\}^m$
- THR_COUNT (threshold counter): indicates how many trials have happened. When THR_COUNT reaches THR_MAX, the tag terminates current session and starts a new session. It can report DoS attack optionally.

The database has the structure shown in figure 4. It must prepare the same number of slots as that of all possible H(ID‖C), while H denotes cryptographic hash function, and ‖ denotes concatenation. When ID is constant, there are 2^m numbers of slots that have the same ID. If H(ID‖C)of different ID conflicts, they are 'linked' in the same slot. If m is 10, the database server has to calculate 1024 $(=2^{10})$ hash values in advance. Though it seems to require much computation of database server, hash values are computed after authentication in the idle time. It is possible to use special purpose unit for computing hash values. Actually, the computations spend small of time (about 0.283 second in internal md5 testing, 0.017 second if only hashing), and we have thought it is not a problem at all if a reasonable m is used.

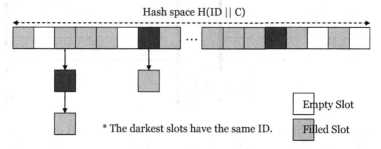

Fig. 4. Example of the database

4.2   Basic Protocol

Our protocol is shown in figure 5 and 6. The protocol solves the security problems referred in section 3.1-3.4.

In order to prevent un-terminated session attack, it must reserve the freshness of response message from tag between sessions. For reserving the freshness of messages, all messages should be generated with secure random nonce at every session. However, if insecure random nonces are used or the attacker can use his number as nonce in authentication session, it can't prevent this attack. Because the tag can't know where the random number included in the request message is from, the random number from a legal reader also isn't trusted. Therefore, only random number from the tag itself is trusted. Consequently, the tag should generate random nonces newly when new session starts.

In order to prevent preemptive locking and stealth bombing attack at once, it needs to handle the session very carefully. Preemptive locking attack is possible because of the session-preemptive feature of the authentication protocol, and stealth bombing attack is mountable because of the session-nonpreemptive feature of the authentication protocol. An authentication protocol cannot be preemtive and non-preemptive at the same time.

Even though the tag receives request message during authentication session, the attacker can not preempt the session if the tag sends the same response message repeatedly until normal termination. We can also frustrate stealth bombing attack using the repeated transmission of the same response message. Note that this method is possible only if random nonces are generated by the tag. By reserving the identicalness of response message from a tag during one session, our protocol can be very robust against preemptive locking and stealth bombing attack. Our protocol provides a strategy that dose not frustrate the attacks, but ignore them.

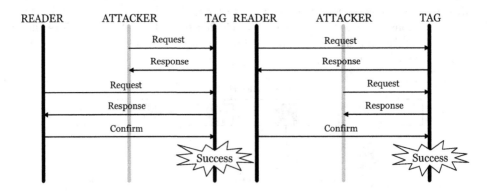

Fig. 5. How to solve preemptive locking and stealth bombing. All response messages have the same value. If a tag can respond with the same response in one session, these two attack can be thwarted.

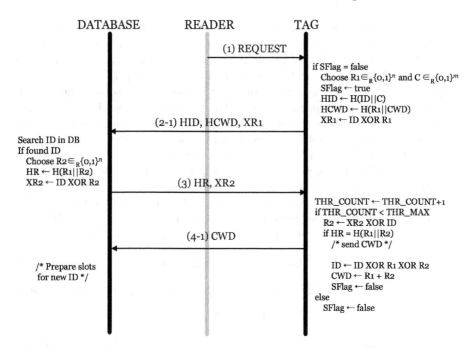

Fig. 6. Proposed protocol

However, to prevent DoS attack using insertion of an invalid confirm message, we need another strategy. To prevent this attack, tag have to wait for a valid confirm message for reserved time. Because it is too expensive to use a timer in RFID tag, threshold counter can be used instead of the timer.

Our protocol illustrated in figure 6 runs as the following:

step 1. When a reader needs tag's data, the reader sends message (1) {REQUEST} to tag

step 2. When a tag receives REQUEST message, it checks SFlag. If SFlag is true, the tag uses R1 and C which are already in memory. Else, the tag chooses random numbers $R1 \in_R \{0,1\}^n$ and $C \in_R \{0,1\}^m$. Also, the tag sets SFlag to true. And then, the tag computes and sends message (2-1) {HID←H(ID∥C), HCWD←H(R1∥R3∥CWD),XR1←ID⊕R1} to the database server through the reader.

step 3. Using HID of message (2-1), the server can get candidates for ID. Also the server gets candidates for R1 by computing XR1⊕(candidates for ID). The server can find a real ID of the tag by checking whether HCWD is the same as H((candidates for R1)∥R3∥(candidates for CWD)) or not.

step 4. If the server can't find any satisfied ID in previous step, the server regards this message as an attack and ignores it. When the server finds ID, the server chooses another random nonce R2. And then, the server generates and sends message (3) {HR←H(R1∥R2),XR2←ID⊕R2} to the tag.

step 5. If the tag receives message (3), it increases THR_COUNT. Then the tag checks whether H(R1∥(ID⊕XR2)) is the same as HR or not if only if THR_COUNT < THR_MAX. If it matches, the tag has to send message (4-1) {CWD} to the database server and continues step 7.

step 6. If the tag has some reason for reject or message (3) is not valid, the tag should wait another messages until THR_COUNT expires. If THR_COUNT reaches THR_MAX (or expires), the tag sets SFlag to false, ignores all next messages and waits until another reader opens a new session.

step 7. If the database server receives correct message (4-1), the server changes its ID and CWD by new ones. Otherwise, the server consider previous message (2-1) to replay attack and halts the process. If the tag has no error to send message (4-1), the tag changes its ID and CWD, and sets SFlag to false. Also the server must prepare hash space for all possible H(ID∥C). Here, CWD consists of R1 and R2 in the previous session. That is,

$$\mathrm{CWD}_i = (\mathrm{R1}_{i-1} + \mathrm{R2}_{i-1}) \bmod 2^n$$

where i denotes the current session and $i-1$ denotes previous session. Also

$$\mathrm{ID}_i = \mathrm{ID}_{i-1} \oplus \mathrm{R1}_{i-1} \oplus \mathrm{R2}_{i-1}$$

In order to find ID from H(ID∥C) in (2-1) message, the database server must prepare all possible slots. If |ID| denotes the number of tags, the number of slots is $2^m \times |\mathrm{ID}|$. If all possible candidates of H(ID∥C) are distributed informly and 2^n is larger then $2^m \times |\mathrm{ID}|$, there is no collision in the same slot. If the system has some collisions, it might have a few liked slots. So, it is good to system to have a number of tags smaller than $2^{(n-m)}$. If the number of tags is about 4,294,967,296(=2^{32}) in system, 42 is enough for n and 10 is enough for m for structure efficiency of the database. But it is strongly recommended to use ID and C with large length for security.

4.3 Security Analysis

Against location tracing: When the attacker wants to trace target tag, he can use message (2-1) and (3) because both message (1) and (4) do not have any information. However, if he can't find any collision pair from hashed messages, he can only use ID ⊕ R1 and ID ⊕ R2 to get clues about the tag. Because he can only get a result of operation ⊕ with these fields, he will only know (ID⊕R1)⊕(ID⊕R2)=R1⊕R2. Since these R1 and R2 are generated by the tag and the server newly at each session, R1⊕R2 also can't be a clue for tracing, even though he can observe all authentication messages.

Also, the attacker can't estimate the messages for next session. Because he knows R1⊕R2, and also knows CWD of tag for next session. However, only with CWD, an attacker cannot trace the location because CWD is masked with R1 such as H(R1∥CWD). An attacker might trace the tag using all possible values of H(ID∥C), but he requires ID which is not exposed to him.

Against spoofing attack: If an attacker wants to spoof the database server or tag for asynchronism between the server and tag, he must generate message (2-1) or (3) correctly. (This asynchronism enables an attacker to mount a kind of DoS attack.) An attacker can generate a valid $H(R1\|CWD)$ only with a probability of 2^{-n} because he must guess a valid ID. In the other hand, an attacker can choose R1 and ID in order to attack against unspecific tag. However, the probability of guessing CWD correctly is 2^{-n}. Since a probability that the guessed ID is found in the database is $|ID| \times 2^{-n}$, he can succeed in this attack having a probability of $|ID| \times 2^{-2n}$.

An attacker can spoof a tag by sending illegal message (3). However, he can generate a valid $H(R1\|R2)$ for $ID \oplus R2$ only with a probability of 2^{-n} since he can extract a valid R1 from $ID \oplus R1$ of message (2-1) if he can guess a valid ID.

Against denial of service attack: DoS attack against a tag was described in section 3.2 and 3.3. Also we explained how to solve these problems at once in section 4.2 and our protocol works according to those principles. Using random nonces from tag and threshold counter for attack trials, the protocol has immunity against the DoS attack. Assuming an attacker cannot modify messages, he has to send or insert messages to mount DoS attack. Even though the attacker tries to do preemptive locking or stealth bombing attack by sending message (1), he cannot succeed because tag answers the request of the attacker using the same response message (2). Also he cannot guess correctly ID, R1 or C from message (2).

When the database server searches ID of the tag, it does not need to hash because all possible values of $H(ID\|C)$ are pre-computed already. Using this strategy, an attacker can't make the database server compute hash value in any cases. Note that the only messages which the attacker can send to the database server are message (2-1) and (4-1). The amount of burden that the database server experiences from one attack message is just as much as one searching of ID space. Thus, even if the attacker can insert or replay message (2-1) and (4-1) in the session, the database server will ignore that message.

When the tag waits message (3), the attacker can send invalid messages to tag for making authentication fail. However, the tag increases THR_COUNT against these messages, and finally authentication will succeed if a valid message arrives before THR_COUNT expires. In the other side, the server must send message (3) in time.

4.4   Alternative Protocol

We propose another protocol, which is a slight modification of our original protocol. The database server authenticates the tag first in the original protocol. But, in the alternative protocol, the tag authenticates the database server first. After message (2-2) is sent, the server makes candidates of ID, and tries to be authenticated by sending message (3) several times until it finds a valid ID. After the tag receives a valid message that includes valid ID and R2, it sends CWD to the server in plaintext, and changes its ID by $ID \oplus R1 \oplus R2$. If a valid CWD arrive, the database

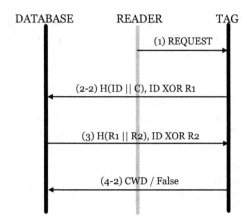

Fig. 7. Alternative Protocol

server will changes ID to ID⊕R1⊕R2 and CWD to R1+R2. Even though R1⊕R2 is obtained easily by observing previous session, it is hard to extract R1+R2 from R1⊕R2. So, the attacker can't get any clues for tracing target tag.

Even though alternative protocol might require slightly more message in average than the original one, it needs only two hash computations.

5   Conclusion

Even if there are many authentication protocols for RFID system, only a few protocols support location privacy. Because of the tag's hardware limitation, these protocols suffer from many security threats, especially from DoS attack.

We established threat model for RFID system and explained some special attack for general authentication protocol. In order to solve these problems, we suggested two strategies: keeping the nonce identical during a session, and threshold counter. With these schemes, we proposed a strong authentication protocol against DoS attack supporting location privacy.

Finally, we checked the strength of our protocol against three categories of attacks, and we concluded that our protocol has reasonable security strength. In addition, we introduced alternative protocol that reduces one more hash operation.

References

1. Dirk Henrici and Paul Müller : Hash-based Enhancement of Location Privacy for Radio-Frequency Identification Device using Verying Identifiers. University of Kaiserslautern, Germany, Workshop on Pervasive Computing and Communications Security - PerSec2004, pp. 149-153, IEEE, 2004
2. Miyako Ohkubo, Koutarou Suzuki and Shingo Kinoshita : Cryptographic Approach to 'Privacy-Friendly' Tags. NTT Laboratories, Japan, RFID Privacy Workshop MIT, 2003

3. István Vajda and Levente Buttyán : Lightweight Authentication Protocols for Low-Cost RFID tags. Budapest University of Technology and Economics, Hungary, 2003
4. Ari Juels : Minimalist Cryptography for Low-Cost RFID Tags. RSA Laboratories, USA
5. Ari Juels : Yoking-Proofs for RFID Tags. RSA Laboratories, USA
6. Ari Juels, Ronald L. Rivest and Michael Szydlo : The Blocker Tag:Selective Blocking of RFID Tag for Consumer Privacy. RSA Laboratories, USA
7. Philippe Golle, Markus Jakobsson, Ari Juels, and Paul Syverson : Universal Re-encryption for Mixnets. RSA Laboratories, USA
8. Stephen J. Engberg, Morten B. Harning and Christian Damsgaard Jensen : Zero-knowledge Device Authentication: Privacy & Security Enhanced RFID preserving Business Value and Consumer Convenience. Privacy, Security and Trust 2004 - PST2004, EU Smarttag Workshop, 2004
9. Martin Feldhofer : A Propsal for an Authentication Protocol in a Security Layer for RFID Smart Tags. Institute for Applie Information Processing and Communications (IAIK), Graz University of Technology, Austria
10. Gildas Avoine and Philippe Oechslim : RFID Traceability: A Multilayer Problem. École Polytechnique Fédérale de Lausanne (EPFL), Switzerland, Financial Cryptography - FC'05, LNCS, Springer, 2005

Location Privacy in Bluetooth

Ford-Long Wong and Frank Stajano

University of Cambridge,
Computer Laboratory

Abstract. We discuss ways to enhance the location privacy of Bluetooth. The principal weakness of Bluetooth with respect to location privacy lies in its disclosure of a device's permanent identifier, which makes location tracking easy. Bluetooth's permanent identifier is often disclosed and it is also tightly integrated into lower layers of the Bluetooth stack, and hence susceptible to leakage. We survey known location privacy attacks against Bluetooth, generalize a lesser-known attack, and describe and quantify a more novel attack. The second of these attacks, which recovers a 28-bit identifier via the device's frequency hop pattern, requires just a few packets and is practicable. Based on a realistic usage scenario, we develop an enhanced privacy framework with stronger unlinkability, using protected stateful pseudonyms and simple primitives.

1 Introduction

1.1 Wireless Devices

Ubiquitous gadgets have been steadily proliferating, posing an increasing threat to personal information privacy. The types of ubiquitous devices may be simplistically arranged on a spectrum according to their intended pervasiveness. On one end we would have the personal cellphone, which we may find one apiece for each individual person, where each cellphone is identifiable and traceable by its network. At the other end, you may find by the hundreds RFIDs—passive radio tags returning 128-bit unique IDs. Privacy solutions for cellphones include the use of network-issued temporary pseudonyms - the 'TMSI', and ways to manage these [1, 2]. Solutions for RFID privacy include 'killing' the tag upon purchase of the attached item, or enclosing it in a mesh, or changing its ID by 're-encrypting' with an external agent, etc. We propose that short-range ad-hoc wireless technologies, such as Bluetooth, which lie in the middle of the spectrum among ubiquitous devices, lend themselves to a different solution framework. Bluetooth devices have finally appeared in large numbers in the past two years after initial problems, and have gained market acceptance and general user familarity. Improving upon a tested and well-received technology may be less painful than designing from ground-up a completely new solution. This article is an attempt to work towards a more refined privacy solution framework for Bluetooth.

R. Molva, G. Tsudik, and D. Westhoff (Eds.): ESAS 2005, LNCS 3813, pp. 176–188, 2005.
© Springer-Verlag Berlin Heidelberg 2005

1.2   Location Privacy

There are different components to privacy. The Common Criteria [3] analyzes privacy into anonymity, pseudonymity, unlinkability and unobservability. Anonymity deals with whether a subject may use a resource without disclosing the user identity. Pseudonymity makes a user accountable for the use, without disclosing his identity, by providing an alias. Unlinkability ensures that a user may make multiple uses of resources or services without others being able to link these uses together. It attempts to obscure the relations between actions by the same user. Unobservability ensures that a user may use a resource without third parties being able to observe that it is being used. For example, a broadcast obscures from third parties who actually received and used that information.

In location privacy, we are concerned with a particular type of privacy, which has been defined as 'the ability to prevent other parties from learning one's current or past location' [4], and it is a relatively new issue in privacy.

1.3   Structure of Paper

In Sections 2 and 3, we cover attacks on the current system. Our location privacy goals are outlined in Section 4. Proposals to assure location privacy are described in Sections 5 and 6.

2   Vulnerabilities

Current well-known authentication weaknesses in Bluetooth could be relatively easily resolved by recourse to asymmetric key establishment techniques [5, 6] at the cost of slightly increased computation. These enhancements would defeat even a strong adversary, by which we mean one which is omnipresent, has significant computational resources, and is able to mount active attacks.

In comparison, it is generally difficult to secure privacy, including location privacy. Awareness to Bluetooth's vulnerabilities in this area was first raised by Jakobsson and Wetzel [7]. Each Bluetooth device is identified by a unique permanent 48-bit Bluetooth Device Address (BD_ADDR). As Bluetooth is usually attached onto personal devices, the detection of a particular BD_ADDR in the neighbourhood would suggest that a particular human operator is nearby. That individual may even be carrying multiple Bluetooth devices and, if such a cluster of BD_ADDRs is detected, it is highly probable that the individual is nearby.

Furthermore, the device's BD_ADDR is used as an input into many procedures in Bluetooth. It is deeply entangled into certain parts of the protocol stack, and it is difficult to engineer it away easily, showing the general difficulty of providing security as an afterthought.

We will provide an overview of aspects of the Bluetooth radio and baseband layers, which are cause for privacy concerns. They can be summarized into:

1. problems of discoverability
2. problems of the non-discoverable mode
3. disclosure of the identity in certain packets

4. derivation of the access code from the identity
5. derivation of the frequency hop set from the identity

We provide a survey of the first three problems, which are well-known. The last two problems have been raised before partially, but we analyze and quantify more fully the risks involved.

2.1   Problems of Discoverability

The purpose of discovery is to allow one to find devices which one has not encountered before. The inquiry scan mode is also known as discoverable mode. The discovering (or inquirying) device sends ID packets, which contain just an access code — either the General Inquiry Access Code (GIAC) or a Dedicated Inquiry Access Code (DIAC), and according to the requisite inquiry hop sequence. A device in the inquiry scan mode will respond to inquiries with a Frequency Hop Synchronisation (FHS) packet, disclosing its own BD_ADDR and CLKN (native clock). The response is not immediate, but is on receipt of the next packet, so as to avoid collision with other slaves.

Essentially, the discovery process enables a hitherto stranger device to be found after at most tens of seconds, and from the privacy perspective the real identity is unfortunately disclosed when a device is discoverable. Keeping devices constantly discoverable is clearly a privacy risk. It is advisable to turn off discoverability whenever it is not needed.

2.2   Problems of Non-discoverable Mode

Devices which are set to 'non-discoverable' are nevertheless responsive to some degree. If they are set to 'connectable', they can still be detected, due to privacy weaknesses in the page and page scan states. During page, the master device will page for another device using an ID packet containing a Device Access Code (DAC) derived from the Lower Address Part (LAP) of the latter's BD_ADDR. The hopping at the physical layer during page is similar to the case for inquiry. The hop sequence is derived from the DAC, instead of the GIAC or some DIAC, together with the estimated clock (CLKE) of the paged device. When the slave detects the page message containing its own DAC, it will reply with an ID packet containing the same DAC. After that, the master will transmit a FHS packet.

Thus, a slave device set to non-discoverable and connectable will not respond to inquiry messages, but it will respond to page messages containing its permanent DAC. Devices which have previously encountered this device and have a record of its BD_ADDR and/or DAC can still page for it successfully if the device is within radio range. If its BD_ADDR is not known, the discovering devices can conduct a brute-force search of the BD_ADDR range, or more precisely, the 24-bit LAP range. The only means of protection against being tracked this way possible under the current specification are to either turn off Bluetooth, or to switch to non-connectable mode if such fine-grained control is supported on the particular device, and lastly, to reduce occurrences of pairing to a minimum so as avoid over-exposing the device's BD_ADDR.

2.3   Disclosure of Identity in FHS Packets

The FHS is a special control packet. The entire BD_ADDR of the sender, comprising the Lower Address Part (LAP), Upper Address Part (UAP), and Nonsignificant Address Part (NAP), are disclosed in the FHS packet, together with the highest 26 bits of its 28-bit native clock CLKN. The FHS packet is sent on two occasions: by a slave device in inquiry scan mode responding to an inquiry; and by a master device in page mode responding in turn after a slave in page scan mode has responded to the page. The device's identity is hence revealed to the opposite party and to any eavesdropper who is monitoring the spectrum.

2.4   Baseband Access Code Derived from Identity

The derivation of the Channel Access Code (CAC) from the master device's LAP had been recognized by Jakobsson et al [7] as a privacy risk, because the LAP can be reverse-engineered. We generalize further that the derivation of not just the CAC but also the DAC from the LAP carry privacy risks.

The access code of a Bluetooth packet is either one of three types. The CAC is used during the Connected state, the DAC is used during page and page scan, and the Inquiry Access Code (IAC) is used during inquiry and inquiry scan. The sync word is a 64-bit code derived from the LAP of a BD_ADDR. In the CAC, the sync word is derived from the master device's LAP; in the DAC, the LAP of the paged slave unit is used; and in the IAC, either the single reserved LAP is used or else certain dedicated IACs are used. The inquiry state is of less interest for privacy because the IACs being correlated for are not too device-specific.

The attacker only needs to compute once a dictionary of 2^{24} (ie. 16.7 million) LAP entries and their corresponding 64-bit sync words. As raised in [7], when the attacker detects a CAC, he can perform a table lookup and learn the master device's LAP. For completeness, we further raise that when this attacker detects a DAC sent by a paging master and a responding slave, he can perform a table lookup using the same pre-computed dictionary, and learn the slave device's LAP. As such, the slave device, non-discoverable but connectable, also faces location privacy risks. Note that a particular LAP is not unique, though collisions would be rare. The remaining address bits — the 24 bits of the UAP and NAP which constitute the 'company_id', do not span the entire 24 bits of entropy — the allocated numbers are published by IEEE Standards, and as of Jan 2005, there were only around 2^{13} issued numbers.

2.5   Hop-Set Derived from Identity

Jakobsson et al [7] observed that since the hop sequence in a connected piconet is a function of the master device's BD_ADDR and CLKN, and thus if one can capture a FHS packet sent by the master, the hop sequence can be trivially calculated. We investigated the reverse attack—the more difficult one of how to recover the master device's address by tracking the frequency hopping pattern if we failed to capture the master's FHS packet, and we found that collecting just 6 packets, along with other information, is adequate. This attack produces 28 bits of address — 4 more bits than attacking the access code.

Bluetooth uses frequency-hopping mainly to mitigate environmental interference, and to reduce collisions among different piconets. There are five types of hopping sequences for the 79-hop system, one type each for the inquiry, inquiry response, page, page response and connected states. Each of these sequences is determined by the 24-bit LAP and the lower 4 bits of the UAP, of the relevant device's BD_ADDR, and its clock. The choice of device address used here is identical to that used to compute the access codes for the different states.

Thus 28 bits of LAP/UAP and 27 bits of the clock go into the hop selection box at any one time to choose one frequency. This function is fully documented in the specification and is strictly surjective. In the connected state, the output selects one of 79 frequencies, corresponding to an approximate 6-bit range. Based on a reasonable assumption of a uniform distribution, thus for the same clock offset, roughly 2^{22} LAP/UAP values would result in the same frequency.

We can carry out the following attack. Capture a first packet and form a tuple of the clock and frequency. Do a brute-force search and narrow the set of 2^{28} LAP/UAP values into a set of about 2^{22} possibilities. Collect another packet and obtain the tuple. Assuming uniform distribution, we can narrow further to a set of 2^{16} possibilities. Continuing in a similar way, just 6 packets in total are required to determine a unique 28-bit LAP/UAP with a probability calculated at 99+%. This is described in Appendix A. The overall work factor is on the order of 2^{28}. With so few packets required for successful attack, the attacker may simply listen at a fixed frequency for it to be re-visited, instead of scanning the entire band. We have to add a caveat that, since the clock setting at each packet is required, determining the master device's clock setting initially without recourse to capturing its FHS packet would entail an indirect route of obtaining a LMP packet containing the slave's clock offset relative to the master's, and inquiring the slave (which needs to be discoverable) to learn the slave's clock.

This novel attack shows that even if a master device is non-discoverable and non-connectable, its hop pattern in a connected state and a discoverable slave could betray its identity.

Bluetooth was not expressly designed to be resistant to interception and deliberate narrowband jamming, unlike, for example, military tactical communications. Our interest with the frequency hop in Bluetooth is on the anonymity issues rather than availability. By resource-sharing the radio access via different clock offsets and public long-term identifiers, frequency hopping achieves equitable allocation of the spectrum and reduces collisions, but it hurts privacy. To improve privacy, the options are: either to disentangle the identifier from the time-frequency allocation, thereby requiring a re-design of the radio layer; or else to just de-link the identifier from the long-term identity, which is simpler.

3   Adversary Types

We identify two classes of adversaries, in ascending order of capability to compromise the privacy of the Bluetooth device.

The first class of attackers use commercial Bluetooth devices that can inquire and page as usual, and can therefore find any discoverable Bluetooth devices, as described in Section 2.1. The attacking range may be extended by directional antennae, a concept well-known to EM/RF engineers. For example, a 18 dBi yagi antenna can boost the 100 m Class I Bluetooth range to around 900 m, and a 24 dBi antenna to 1.6 km, assuming low RF losses at the joints, though such antennae are large and obtrusive. Within this class of attackers, we can distinguish a slightly more sophisticated sub-class, who can conduct brute-force searches of the BD_ADDR space, or rather, the LAP space, so as to find connectable victim devices, as described in Section in 2.2. Such proof-of-concept code has been released [8], though it is estimated to require around 11 hours to conduct a complete search of the space, using 127 devices working in parallel. We have developed our own version of this attack using a shell script, the open-source BlueZ stack, and an ordinary Bluetooth dongle.

A second class of attackers uses radio receivers, or modified Bluetooth devices, which are not constrained to frequency hop. The first sub-class can listen on one selected channel continuously for all types of messages in the inquiry, inquiry scan, page, page scan, and connected state hops. If this attacker sees a CAC or DAC, he can carry out his table lookup privacy attack, as described in Section 2.4. If he sees a FHS packet, then he has learnt the full BD_ADDR, as described in Section 2.3. He can also derive the master's identity by knowing at which clock offsets a particular hop frequency is re-visited, and by probing a discoverable slave, as described in Section 2.5. Another more powerful sub-class is capable of listening on the entire 2.4 GHz band simultaneously. This attacker is less likely to miss any packets, and is more effective than the first sub-class in determining the CLKN of the target master device for the attack in Section 2.5. Attacking the access code is less costly than attacking the frequency hop pattern though. The first sub-class of attacks can be readily demonstrated with today's Bluetooth protocol analyzers, such as the Frontline-Tektronix BPA-100 and 105.

We distinguish between hardware, and do not distinguish between the cryptographic capability among the classes, because programs which do such computations can be commoditised easily and can run on generic PCs. The first category of adversaries are able to successfully compromise the privacy of today's Bluetooth devices easily, unless tight discipline is maintained over the use of the discoverable mode and connectable mode. The second category of attackers is able to compromise the privacy of Bluetooth devices even when their victims maintain tighter discipline over discoverability and connectability, and whenever devices are transmitting in a connected state. The overall efficacy of location privacy attacks also depends on the pervasiveness (and investment) of the attackers, and how effectively they can correlate and fuse information obtained by their various spatially distributed sensors to continuously track the location of their victims.

4   Location Privacy Goals

The current specification of Bluetooth does not support strong location privacy. Before we go into the detailed technical mechanisms, we need to define the usage scenarios for this short-range wireless connectivity technology. Then we will articulate the privacy goals which take into account the usage.

Bluetooth-equipped devices tend to talk to other personal devices, and less with fixed immobile network infrastructure. The interaction is mostly peer-to-peer. Users of Bluetooth do not seem to require it to have substantial location-awareness for it to work well for cable-replacement. A higher application layer may require location-awareness, but Bluetooth, as a connectivity layer, does not require location-awareness built-in, and can very well lean towards the location-private part of the continuum. These differences make its location privacy requirement different from other technologies which have been analyzed elsewhere, which had assumed a network backbone [4]. We admit that the security interaction of Bluetooth with the location-aware parts, where present, of the host device may merit further study.

On the other hand, devices hosting Bluetooth are rather much smarter than dumb tags such as RFIDs. Bluetooth interactions may be stateful, since session keys need to be established. Identifiers are required for this and cannot be eliminated. This is true for both piconet and scatternet configurations.

Temporary throwaway pseudonyms [9, 5, 10] can be of help. However, these must not be completely stateless, otherwise prior pair-wise relationships and piconet configurations would be quickly lost, and require frequent re-initialization. From the point of view of privacy, the need for a permanent identifier is debatable. Apart from helping manufacturers tell their product lines apart, having hierarchically arranged BD_ADDRs does not appear to do privacy much good.

Spectrum allocation and collision avoidance at the physical layer have been mentioned to have privacy implications. A good solution must resolve these.

While we have discussed exclusively about Bluetooth, in practice some other protocol is sometimes tunneled over Bluetooth. One important issue for anonymity is that the different protocols must carry out proper de-identification between them and be stateless. For example, if TCP/IP is tunneled over Bluetooth, the BD_ADDR should be de-linked from the IP address. However, we will consider this as outside the scope of this article.

Thus we require a privacy framework which provides sender and destination anonymity in a mostly peer-to-peer ad-hoc wireless environment. Pseudonyms may be used, and unlinkability between pseudonyms should be provided. The solution should account for cases in which the wireless personal area network stays in a static configuration, and for cases where state needs to be kept between two paired devices over different sessions due to the inconvenience of establishing a new session key. Unobservability should be provided. If the premises underlying the usage scenario evolve, the privacy framework needs to change too. The means to establish strong pair-wise keys is assumed to exist [5, 6]: this is a non-goal.

5   Problems of Pseudonyms and Permanent Identifiers

As the identifier BD_ADDR is tightly integrated in the protocol and is used in many computations, it cannot be easily discarded. Throwaway pseudonymous 'BD_addr_actives' were proposed by Gehrmann and Nyberg [5] to be used within an anonymity mode. Using frequently changing pseudonyms would improve the unlinkability between actions by the same actual principal, and also protect the permanent BD_ADDR, which the device still retains, from disclosure to a casual observer. Using pseudonymous BD_addr_actives this way also allow the original design of the access code and frequency hop to be essentially retained.

However, that proposal has three privacy weakness. The first is that the real identity, the BD_ADDR, is being used and may be disclosed to any device with which one has paired previously, though the identity is protected against other casual observers. Thus, adversaries can link different actions to the same actual principal if they can pair with this device, no matter what its particular BD_addr_active is at the instant. This is not an ideal privacy quality to possess, as policy-wise it should not automatically be assumed that all devices which have paired with one's own are not adversarial with respect to one's privacy.

A second weakness is with regards to the usage of BD_addr_alias, which is another 'BD_ADDR-like' identifier, established by two devices after they have paired, to signify the pairing in their respective database. For example, a BD_addr_alias would in Alice's database serve as an alias signifying Bob to Alice, and in Bob's database as an alias signifying Alice to Bob. In Alice's database there would be a tuple containing this BD_addr_alias and Bob's real BD_ADDR. In one mode, after Alice pages Bob, and before authentication takes place, Alice would send a packet containing this BD_addr_alias to Bob in an attempt to find out if they have paired before. Bob will now look up this alias in his database to find Alice's BD_addr, and respond accordingly. The problem with this usage is as follows: if Alice pages for Bob, but this is intercepted by an adversary Eve, and Eve receives the BD_addr_alias sent by Alice, while Eve will fail the test, Eve would be able to page Bob later using the alias, and thence be able to probe whether Bob has previously paired with Alice. The observability of transactions between Alice and Bob could thus be compromised offline.

A third privacy weakness is related to the second. An adversary who observes the same pairwise BD_addr_alias transmitted can deduce that the same two devices may be communicating again. There are other caveats concerned with the use of temporary pseudonyms, which we would discuss. One of the most germane ones is that if a device could continually be tracked, even as it changes its pseudonym, that could still be linked to the previous one.

We propose an enhanced anonymity mode, also using pseudonyms, which would attempt to address these three said problems, while recognizing that pairings may be stateful. We emphasize that this mode by itself will not resolve all privacy risks; a policy which requires discoverability and connectability to be turned off most of the time must be applied.

6   Proposed Solution: Protected Stateful Pseudonyms

6.1   Inquiry and Inquiry Scan

For device discovery, we keep to the Gehrmann and Nyberg proposal [5], where the inquiry and inquiry scan states are left as according to the original specification, with the change that the identifier returned at inquiry scan is the slave's BD_addr_active instead of its BD_ADDR.

As a matter of strong privacy policy to counter tracking, we recommend that a device's discoverability should be turned off whenever it is not required.

6.2   Page and Page Scan

The Gehrmann and Nyberg proposal featured two paging situations. One situation is where a master pages a slave based on the latter's current BD_ADDR_active. The second situation is where a master pages a slave based on the latter's long-term BD_ADDR, which is useful for previously paired units. The second situation allows pairings to be remembered, but has the unfortunate weakness of leaking the BD_ADDR of the slave being paged, hence compromising linkability as well.

We prefer that the long-term BD_ADDR never be leaked. We hence propose a somewhat different second situation, in which a master would attempt to page a slave using modified ID packets derived from the previous BD_ADDR_actives which the master and the slave had used to pair. These packets cryptographically protect the addresses from casual sniffing. The formats of the packets and the required protocol are as described in the following section. We believe that it is more private to have done pairing with the pseudonyms than with the long-term identifiers. This is not too difficult to support, as Bluetooth pairing is already based on a shared password rather than on permanent identifiers. It can be decided by policy settings how soon to expire pairings, as well as how soon a device expects a paired device to have changed pseudonyms.

As a policy setting, we recommend that a device's connectability be turned off whenever the owner does not expect connection requests to be received.

6.3   Protected Pseudonyms

This protocol (Fig. 1), designed for our second situation described above, attempts to protect past pseudonyms from all third parties. We modify the ID packet from the original Bluetooth specification. We now use three ID packets, denoted by ID1, ID2 and ID3. The relevant past pseudonyms of Alice and Bob are denoted by I_A and I_B. H is a hash function, R_1, R_2 and R_3 are random nonces, and K_{AB} is the shared link key formed by Alice and Bob previously. The three-way handshake is essential. Say, Alice intends to page for Bob. On verifying correctly the ID2 packet, Alice will have the assurance that Bob knows his previous pseudonym, her previous pseudonym, and their shared key. On verifying correctly the ID3 packet, Bob has the assurance that Alice knows these same three things.

#	Alice	Bob
1	Chooses random R_1	
2	$H_1 = H(I_B|R_1|K_{AB})$	
3	$- ID1 : (R_1 \mid H_1) \rightarrow$	
4		Verifies H_1
5		Chooses random R_2
6		$H_2 = H(I_A|R_1|R_2|K_{AB})$
7	$\leftarrow ID2 : (R_2 \mid H_2)-$	
8	Verifies H_2	
9	Chooses random R_3	
10	$H_3 = H(I_B|I_A|R_1|R_2|R_3|K_{AB})$	
11	$- ID3 : (R_3 \mid H_3) \rightarrow$	
12		Verifies H_3

Fig. 1. Protected Stateful Pseudonyms

Alice keeps a database of tuples each containing her temporary pseudonym, the pseudonym of the other party, and the shared link key. Bob keeps a similar database. Alice wants to page for Bob. She selects a random nonce R_1, computes the hash H_1, and sends an ID1 packet. The hash in the ID1 packet hides the past pseudonym of Bob. Bob would compute and verify the expected hash in the ID1 packet using his list of the paired devices' pseudonyms and their associated link keys with the nonce. When he successfully finds a match, he chooses a random nonce R_2, computes H_2, and responds with the ID2 packet. The hashes are inexpensive operations, thus the parties can do these easily. As Bob generates nonce R_2 randomly, he can be sure that his challenge to Alice is fresh. Alice, on receiving the ID2 packet, will verify the hash. If there is a match, Alice will generate a nonce R_3, compute the hash H_3, and reply with the ID3 packet. Bob will verify the hash on receipt of the $ID3$ packet. After the protocol runs successfully, both parties can proceed to carry out mutual authentication as usual. The security of the protocol depends on the randomness of the nonces, the irreversibility of the hash function, and the secrecy of the shared link key.

A naive replay attack — the second weakness mentioned in Section 5 — incarnated here as an adversary capturing an ID1 packet previously sent by Alice and received by Bob, and replaying it, would be defeated, because Bob checks for freshness of R_1 and R_3, and Alice checks for freshness of R_2

In another conceivable and more sophisticated attack, an adversary Eve intercepts an ID1 packet and prevents it from reaching Bob, but replays it later to Bob. Such an ID1 packet will pass Bob's R_1 freshness test. However, Bob now sends an ID2 packet with a fresh R_2. We can set a policy whereby uncompleted handshakes would raise an alarm at Bob's end, to alert Bob of the possibility of an intruder, so unless Eve next responds with a correctly formed ID3 packet, Bob would receive an alert. The 3-way handshake is essential for mitigating such an attack. Over at Eve's end, on her receipt of Bob's ID2 packet, Eve may suspect that Alice and Bob had paired previously, but she retains some doubt, because of possible collisions among the hashes.

The protocol is not resistant to an online relay attack — in which Eve would position herself between two widely geographically separated victims — because the protocol does not incorporate any distance-bounding algorithm.

We leave it open whether the length of the ID1, ID2 and ID3 packets need to be equivalent to the DAC length of 68 bits. If they are also 68-bit, especially the ID1 packet, then it helps to obscure the fact from simplistic traffic analysis that Alice is paging for a old pseudonym of Bob, in which case the random nonce would take up, say, 34 bits, and the hash the other 34 bits. Or else these packets can extend up to the length of 160 bits plus a suitable length of a nonce.

The proposed protocol provides good scalability in remembering and responding to past pairings, while not leaking the permanent BD_ADDR nor previous pseudonyms unnecessarily. Bob keeps changing pseudonyms, yet remains able to respond to some previous pseudonym of his which Alice has paired with, as he has kept a history of his pseudonyms, whose ages are set by policy.

6.4   Physical Layer

We have described in Section 2 that parts of the device address can be recovered from the access code and the frequency hop pattern. This privacy risk can be resolved by using changing pseudonyms. A more complex possible solution would be to modify the physical layer. The frequency hop pattern can, for example, be initialized from other parameters instead of a device's identifier and its clock. Another alternative solution is to use direct-sequence (DS) spread spectrum instead of frequency hopping, so that different DS sequences use different pseudonymous identifiers. But the cost of DS is generally considered higher.

6.5   Triggers for Pseudonym Change

We propose several triggering mechanisms to change pseudonyms. It is well-known that if a device can be continuously tracked, such as when it is discoverable and is the only device in a locality, then even a change of identifier would not prevent linkability. Discoverability ought to be turned off during pseudonym change. We suggest a sub-state in the anonymity mode in which the device is ready to change pseudonyms. A change may be triggered by any of several events. Firstly, it may be brought about by the owner's manual action. Secondly, it can be automatically changed at random time intervals. Thirdly, the pseudonym be changed when a certain threshold large number of discoverable devices are detected in an inquiry sweep. The rationale is that it would be easy to 'blend in with the crowd' and anonymise oneself. This method should be carefully applied because an attacker can spoof the presence of a large number of devices. [1] It uses the concept of a mix zone [4], the difference being that here, pseudonym change is handled by the devices themselves instead of a network infrastructure.

---

[1] However, the attacker would not reduce the anonymity of the victim by forcing a pseudonym change — the only effect would be to make the victim believe he is more anonymous that he actually is, which might perhaps lead him to lower his guard.

6.6 Further Issues

New pseudonyms must be randomly generated, and one solution is by hashing some counter. Also, the 28-bit 3.2 kHz Bluetooth device clock, which has a cycle of 23.3 hours, of which the highest 26 bits are disclosed (— or a 1.25 ms resolution), must be randomly re-adjusted on a pseudonym change, to prevent an adversary from linking pseudonyms to a clock, even accounting for the clock drift. Bluetooth uses a 'friendly name', which is a human-readable name to tag devices during device discovery and to help manage the list of paired devices locally. To reconcile privacy with usability, we propose the following: the field should be left empty or not transmitted during device discovery, but the user could be allowed to locally tag his list of paired devices with 'friendly names' of his choice to help him better distinguish the devices than through hexadecimals.

Certain RF attacks attempt to pinpoint the location of devices by measurements of irradiated power, and more sophisticated attacks distinguish RF signatures of individual devices, but these are outside our scope. In our privacy framework, we have not made use of digital certificates, because though these allow strong authentication, they are inimical to anonymity.

7 Conclusions

We have investigated the privacy problems of this pervasive wireless ad-hoc technology, particularly the leakage of its unique device address. We have surveyed known attacks against its location privacy, expanded a previously raised attack, and quantified another less-studied attack. The last attack requires only several packets and a work factor of 2^{28}. While the basic location privacy problem of using a long-term device address can be resolved by using temporary pseudonyms, an incomplete solution can give rise to linkability.

Based on a plausible usage scenario distinct from other wireless technologies, we propose ways which refine the use of the pseudonyms, so that they are stateful and past device pairings can be remembered according to policy, yet which do not leak past pseudonyms and the long-term device address unnecessarily. We have also described various mechanisms to manage pseudonym change.

Acknowledgement

We are grateful to the anonymous reviewers for their helpful comments.

References

1. D. Kesdogan, H. Federrath, A. Jerichow and A. Pfitzmann. "Location Management Strategies increasing Privacy in Mobile Communication Systems". *Proceedings of the 12th IFIP SEC*, 1996.
2. S. Capkun, J. Hubaux and M. Jakobsson. "Secure and Privacy-Preserving Communication in Hybrid Ad Hoc Networks". *EPFL-IC Technical report IC/2004/10*, Jan 2004.

3. ISO/IEC-15408 (1999). *ISO/IEC-15408 Common Criteria for Information Technology Security Evaluation v2.1*, 1999. `http://csrc.nist.gov/cc`.
4. A. R. Beresford and F. Stajano. "Location privacy in pervasive computing". *IEEE Pervasive Computing*, **3**(1):46–55, 2003.
5. C. Gehrmann and K. Nyberg. "Enhancements to Bluetooth Baseband Security". *Proceedings of Nordsec 2001*, Nov 2001.
6. F.-L. Wong, F. Stajano and J. Clulow. "Repairing the Bluetooth Pairing Protocol". *Thirteenth International Workshop in Security Protocols*, Apr 2005.
7. M. Jakobsson and S. Wetzel. "Security Weaknesses in Bluetooth". *Proceedings of the RSA Conference*, **LNCS 2020**, 2001.
8. O. Whitehouse. "RedFang". 2003. `http://www.atstake.com/`.
9. D. Chaum. "Untraceable Electronic Mail, Return Addresses, and Digital Pseudonyms". *Communications of the ACM*, **24**(2):84–88, Feb 1981.
10. M. Gruteser and D. Grunwald. "Enhancing Location Privacy in Wireless LAN through Disposable Interface Identifiers: A Quantitative Analysis". *First ACM International Workshop on Wireless Mobile Applications and Services on WLAN Hotspots*, 2003.
11. Bluetooth SIG Security Experts Group. *Bluetooth Security White Paper*, **1.0**, April 2002.
12. Bluetooth Special Interest Group. "Bluetooth Specification Volume 1 Part B Baseband Specification". *Specifications of the Bluetooth System*, **1.1**, Feb 2001.
13. Bluetooth Special Interest Group. "Bluetooth Specification Volume 2 Part H Security Specification". *Specification of the Bluetooth System*, **1.2**, Nov 2003.

Appendix A: Recovery of Address Bits from Frequency Hop

The mathematics of the method can be formulated as a binomial distribution. We assume that each of the 79 outputs is equi-probable. We want to find the probability that after k rounds, only one input is left, ie. all of the other $2^{28} - 1$ inputs are discarded at some round. Each one of these remains with probability $(1/79)^k$. Assuming independence of clock values, and independence between the outcomes of different inputs, the probability we seek is

$$(1 + x)^n = (1 - (\frac{1}{79})^k)^{2^{28} - 1}$$

As the exponent is large, the numerical result is difficult to compute. Since x is small with respect to 1, we can do a binomial expansion.

$$(1 + x)^n = 1 + \frac{nx}{1!} + \frac{n(n-1)x^2}{2!} + \cdots$$

For $k = 6$, the first two terms sum to 0.9989. If we approximate $1/79$ to $1/2^6$, the result is 0.9961.

An Advanced Method for Joint Scalar Multiplications on Memory Constraint Devices

Erik Dahmen[1], Katsuyuki Okeya[2], and Tsuyoshi Takagi[3]

[1] Technische Universität Darmstadt, Fachbereich Informatik,
Hochschulstr.10, D-64289 Darmstadt, Germany
dahmen@rbg.informatik.tu-darmstadt.de
[2] Hitachi, Ltd., Systems Development Laboratory,
1099, Ohzenji, Asao-ku, Kawasaki-shi, Kanagawa-ken, 215-0013, Japan
ka-okeya@sdl.hitachi.co.jp
[3] Future University - Hakodate,
116-2 Kamedanakano-cho Hakodate Hokkaido, 041-8655, Japan
takagi@fun.ac.jp

Abstract. One of the most frequent operations in modern cryptosystems is a multi-scalar multiplication with two scalars. Common methods to compute it are the Shamir method and the Interleave method whereas their speed mainly depends on the (joint) Hamming weight of the scalars. To increase the speed, the scalars are usually deployed using some general representation which provides a lower (joint) Hamming weight than the binary representation. However, by using such general representations the precomputation and storing of some points becomes necessary and therefore more memory is required. Probably the most famous method to speed up the Shamir method is the joint sparse form (JSF). The resulting representation has an average joint Hamming weight of $1/2$ and it uses the digits $0, \pm 1$. To compute a multi-scalar multiplication with the JSF, the precomputation of two points is required. While for two precomputed points both the Shamir and the Interleave method provide the same efficiency, until now the Interleave method is faster in any case where more points are precomputed. This paper extends the used digits of the JSF in a natural way, namely we use the digits $0, \pm 1, \pm 3$ which results in the necessity to precompute ten points. We will prove that using the proposed scheme, the average joint Hamming density is reduced to $239/661 \approx 0.3615$. Hence, a multi-scalar multiplication can be computed more than 10% faster, compared to the JSF. Further, our scheme is superior to all known methods using ten precomputed points and is therefore the first method to improve the Shamir method such that it is faster than the Interleave method. Another advantage of the new representation is, that it is generated starting at the most significant bit. More specific, we need to store only up to 5 joint bits of the new representation at a time. Compared to representations which are generated starting at the least significant bit, where we have to store the whole representation, this yields a significant saving of memory.

Keywords: elliptic curve cryptosystem, joint sparse form, left-to-right, multi-scalar multiplication, shamir method.

R. Molva, G. Tsudik, and D. Westhoff (Eds.): ESAS 2005, LNCS 3813, pp. 189–204, 2005.
© Springer-Verlag Berlin Heidelberg 2005

1 Introduction

In our modern society it becomes more and more necessary to communicate and authenticate electronically in a secure way. Because of their mobility and tamper resistance, smart cards are often used for this task. However, since smart cards are quite small, the available memory and computational power is very limited. The Elliptic Curve Cryptosystem (ECC) [Kob87, Mil86] is an efficient cryptosystem, which can attain high security with very short key length. Therefore, the ECC is suitable for implementation on devices where computational power is limited. The basic operation for verifying a signature with the ECC is a so-called multi-scalar multiplication

$$uP + vQ$$

for given scalars u, v and points on the elliptic curve P, Q. The research goal from practical requirement is to efficiently compute this multi-scalar multiplication by minimizing both memory usage and computational costs [Ava02, Gor98, Möl01]. Another example for resource constraint devices where the ECC can be implemented are sensors. Those devices are mainly used to monitor a given physical environment. Different from smart cards, sensors have no external power source, therefore efficiency is not only required to save time and resources, but also to save energy.

Two efficient methods to compute a multi-scalar multiplication are the Shamir method [ElG85] and the Interleave method [Möl01]. The speed of those methods depends on the (joint) Hamming weight of the scalars. If some memory for precomputation is available they can be sped up by deploying a redundant representation of the scalars. Those representations provide a smaller (joint) Hamming weight than the binary representation, but use a larger digit set. The downside is, that the size of the digit set determines the number of points to precompute, and thus a trade-off between memory usage and speed has to be made.

Probably the most popular representation to speed up the Shamir method is the Joint Sparse Form (JSF) proposed by Solinas [Sol01]. The used digits in this representation are $0, \pm 1$ and it requires the precomputation of two points. The resulting average joint Hamming density is $1/2$. While for two precomputed points, the Shamir method and the Interleave method can provide the same efficiency, until now the Interleave method is faster in any case where more points are precomputed.

In this paper we propose a new representation to further speed up the Shamir method. This is achieved by naturally extending the digit set of the JSF and allowing the digits $0, \pm 1, \pm 3$. For those digits, the precomputation of ten points is required. The main idea of the algorithm is to apply a sliding window method with variable width on both scalars. The widths are chosen such that the resulting density of non-zero columns increases from step to step. Also, if certain conditions are satisfied we reuse already converted columns in the proceeding step. We will prove that the average joint Hamming density of the proposed scheme is $239/661 \approx 0.3615$, which is superior to any known method using ten precomputed points. Therefore, our method is the first to improve the Shamir method such that it is faster than the Interleave method. Compared to the JSF, the computation of $uP + vQ$ can be sped up by more than 10%. Another advantage of the proposed scheme is, that it is generated starting at the most significant bit, which is more natural and memory saving in conjunction with the ECC (see Section 2). More specific, we need to scan only up to 6 joint bits of the binary representations of the scalars at once.

The rest of the paper is organized as follows: In Section 2 we give an overview of multi-scalar multiplication. In Section 3 we review several known methods to speed up the computation of $uP + vQ$. In Section 4 the proposed scheme is described and its computational cost is calculated. In Section 5 we compare our scheme to the methods of Section 3 and Section 6 states our conclusion.

2 Preliminaries

2.1 Notations

A scalar d is a positive integer and there are several ways to represent it. The most common one is the uniquely determined binary representation $(d_{n-1}, \ldots, d_0)$, where $d = \sum_{i=0}^{n-1} d_i \cdot 2^i$ and $d_i \in \{0, 1\}, \forall i = 0, \ldots, n-1$. Here, n is the bit length of the representation. Another way is a more general approach. Now we don't restrict the digits to the set $\{0, 1\}$ but to an arbitrary digit set $\mathcal{D}$. We call $(d_{n-1}, \ldots, d_0)$ a $\mathcal{D}$-representation of d, if $d = \sum_{i=0}^{n-1} d_i \cdot 2^i$ and $d_i \in \mathcal{D}, \forall i = 0, \ldots, n-1$. For example, if $\mathcal{D} = \{0, \pm 1\}$ we call the underlying $\mathcal{D}$-representation a signed binary representation. In general, $\mathcal{D}$-representations loose the property of uniqueness.

The Hamming weight (HW) of a $\mathcal{D}$-representation is its number of non-zero entries. The Hamming density (HD) is defined as HW/n. The average Hamming density (AHD) is the expected HD for a random representation with bit length $n \to \infty$. If we consider more than one $\mathcal{D}$-representation simultaneously, we may want to examine non-zero columns rather than non-zero entries. The number of non-zero columns of an arbitrary number of $\mathcal{D}$-representations is given by the joint Hamming weight (JHW). The joint Hamming density (JHD) and the average joint Hamming density (AJHD) are defined in accordance to the HD and the AHD, respectively. For simplicity, we consider only representations with the same bit length n. This can be achieved by padding with zeros to the left.

Let $K = GF(p)$ be a finite field, where $p > 3$ is a prime. Let E be an elliptic curve over K. The elliptic curve E has an abelian group structure with identity element $\mathcal{O}$ called the point of infinity. A point $P \in E$ is represented as $P = (x, y)$. The inverse of point $P = (x, y)$ is equal to $-P = (x, -y)$, hence it can be computed virtually free. For that reason, it is advisable to use a signed binary representation of the scalars [MO90]. Note that this is also true for elliptic curves over different fields, e.g. binary curves. The elliptic curve operations $P + Q$ and $2P$ are denoted by ECADD and ECDBL, respectively, where $P, Q \in E$.

2.2 Multi-scalar Multiplication Algorithms

In this section we explain how the Shamir method and the Interleave method compute $uP + vQ$. This is done in the so-called *evaluation stage*: at first an accumulator X is initialized with the neutral group element $\mathcal{O}$, then the following steps are performed.

Shamir Method _____ Interleave Method _____

for $i = n - 1$ down to 0 **do**
$\quad X \leftarrow ECDBL(X)$
$\quad$ **if** $(u_i, v_i) \neq (0,0)$ **then**
$\quad\quad X \leftarrow ECADD(X, u_iP + v_iQ)$

for $i = n - 1$ down to 0 **do**
$\quad X \leftarrow ECDBL(X)$
$\quad$ **if** $u_i \neq 0$ **then**
$\quad\quad X \leftarrow ECADD(X, u_iP)$
$\quad$ **if** $v_i \neq 0$ **then**
$\quad\quad X \leftarrow ECADD(X, v_iQ)$

After the last iteration X contains the result $uP+vQ$ and is returned. Because both methods frequently use points of the form v_iQ, u_iP and $u_iP + v_iQ$ it is preferable to precompute and store those points. This is done in the *precomputation stage* which is executed prior to the evaluation stage. Note, that since point inversions can be performed online, we don't have to precompute all required points. Typically one uses a symmetric digit set of the form $\mathcal{D} = \{0, \pm1, \ldots, \pm x\}$. In that case only half of all used points have to be precomputed. The Interleave method computes $t_1P, \forall t_1 \in \mathcal{D}_1 : t_1 > 1$ and $t_2Q, \forall t_2 \in \mathcal{D}_2 : t_2 > 1$, where $\mathcal{D}_1$ and $\mathcal{D}_2$ are the digit sets of the scalars u and v, respectively. The Shamir method computes the points $tP, tQ, \forall t \in \mathcal{D} : t > 1$ and $t_1P + t_2Q, \forall t_1, t_2 \in \mathcal{D} : t_1 > 0$, where $\mathcal{D}$ is the digit set of both scalars. Hence, the total number of precomputed points is $(|\mathcal{D}_1| - 3)/2 + (|\mathcal{D}_2| - 3)/2$ and $(|\mathcal{D}| - 1)^2/2 + |\mathcal{D}| - 3$ for the Interleave method and the Shamir method, respectively.

The average speed of both methods is determined by the number of ECDBL and ECADD operations used. The ECDBL operation is performed in each iteration in both methods, i.e. n times. The Shamir method performs an ECADD operation every time a non-zero column is found, therefore the average number of ECADD operations equals n times the AJHD of the scalars. The Interleave method performs an ECADD operation every time a non-zero entry is found in any of the scalars, therefore the average number of ECADD operations equals n times the sum of the AHD of the scalars.

2.3 Left-to-Right vs. Right-to-Left

Now we explain why it is preferable to perform the evaluation starting at the most significant bit, i.e. left-to-right (LtR), rather than the least significant bit, i.e. right-to-left (RtL). Although both the Shamir method and the Interleave method use a LtR evaluation stage, there also exist methods which use a RtL evaluation stage. The main drawback of those methods is that they are very inefficient when used with general $\mathcal{D}$-representations. Namely, in each iteration, they have to perform one ECDBL operation for all points which might be required in the ECADD step. Those points are all precomputed points plus the base points P and Q. On the other hand, the LtR methods always use the same, fixed points for the ECADD step. Therefore, it is possible to speed up this step significantly if those points are represented in affine coordinates [CMO98].

2.4   A Special Signed Binary Representation

Now we introduce a special signed binary representation which is required to generate our proposed representation. This signed binary representation was proposed independently by two parties and is called the "alternating greedy expansion" [GHPT03] or the "mutual opposite form" [OSST04]. Let $(d_{n-1}, \ldots, d_0)$ be the binary representation of an integer d. We define $\mu_i = d_{i-1} - d_i$ for $i = 0, \ldots, n$, where $d_n = d_{-1} = 0$. Since

$$(\mu_n, \ldots, \mu_0) = (d_{n-1}, \ldots, d_0, 0) - (0, d_{n-1}, \ldots, d_0) = 2d - d = d$$

this operation indeed yields a signed representation of d. Note that this representation, from now on called MOF, can be obtained from LtR and from RtL likewise. The MOF of a non-zero scalar satisfies the following properties:

1. The signs of adjacent non-zero bits (without considering zero bits) are opposite.
2. The most non-zero bit and the least non-zero bit are 1 and $\bar{1}$, respectively.

Further, MOF uses the digit set $\mathcal{D} = \{0, \pm 1\}$ and provides a AHD of $1/2$. Also it has been proven that each n-bit integer has a unique representation as $(n+1)$-bit MOF.

3   The Shamir Method vs. the Interleave Method

As we saw in Section 2.2, the number of ECADD operations of the Shamir method and the Interleave method depends on the JHW and the HW of the scalars, respectively. Hence, in order to speed up those methods these numbers have to be decreased. This is achieved by applying a so-called *recoding algorithm* which rewrites the binary representation of the scalars into some $\mathcal{D}$-representation. There are two kinds of recoding algorithms: those which decrease the HW and therefore speed up the Interleave method and those which decrease the JHW and therefore speed up the Shamir method. Also, the direction in which the scalars are recoded is important. In the case of a RtL recoding algorithm the scalars must be recoded in a separate stage prior to the precomputation stage, because we use a LtR evaluation stage. Then it is necessary to store the whole recoded scalars, which requires $O(n)$ bits memory for each scalar. In the case of a LtR recoding algorithm the recoding can be performed "on-the-fly" during the evaluation stage. The advantage is obvious, now we don't have to store the whole recoded scalars, but only a small part which leads to a significant memory saving.

   In Section 2.2 we also saw that the size of the digit set determines the number of points to precompute. However, the size of the digit set also affects the AHD or AJHD of the $\mathcal{D}$-representation produced by a recoding algorithm, but in a non-proportional way. Therefore we face a trade-off between memory usage for the precomputed points and speed for the multi-scalar multiplication.

This section serves two purposes. At first we review several known recoding methods and explain how they speed up the computation of $uP+vQ$. Second, we explain why the optimal choice for the Shamir method is to use representations which require ten precomputed points. Note, that all representations reviewed in this section are uniquely determined and at most one bit longer than the corresponding binary representation (see the respective reference).

Known Methods Using Two Precomputed Points. The most common recoding algorithm to decrease the JHW is the joint sparse form (JSF) proposed by Solinas [Sol01]. Its AJHD is $1/2$ and it uses the digit set $\mathcal{D} = \{0, \pm1\}$. The drawback of the JSF is, that it can only be generated from RtL. A similar method was proposed in [HKPR04]. It uses the same digit set and provides the same AJHD as the JSF, but it can be applied from LtR. The main idea of this algorithm is to apply a LtR sliding window method with different widths on the MOF of both scalars. At first the width $w = 2$ is tested and if no zero column can be generated the width is increased to $w = 3$. The precomputed points for those two methods are $\{P + Q, P - Q\}$.

To decrease the HW of the scalars, there also exists RtL and LtR variants. One RtL method is the famous width-w non adjacent form (wNAF) [Sol00, BSS99, MOC97]. It uses the digit set $\mathcal{D} = \{0, \pm1, \ldots, \pm2^{w-1} - 1\}$ and its AHD is $1/(w + 1)$. Its LtR equivalent is called the width-w mutual opposite form (wMOF) and is generated by applying a width-w sliding window method from LtR on the MOF each scalar separately [OSST04]. Another LtR method, which is directly applied on the binary representation was proposed in [Ava04]. Both these methods also use the same digit set and provide the same AHD as wNAF. In the case of two precomputed points we choose $w = 3$ and precompute $\{3P, 3Q\}$. The resulting AHD of each scalar then is $1/4$. Therefore, the average density of ECADD operations used by the Interleave method is $1/2$, the same as for the Shamir method using the JSF or the scheme described in [HKPR04].

However, two precomputed points require only 80 bytes of memory and since the current smart card technology offers several kbytes of memory, a lot of memory is waisted. If we want to use more memory, the logical step is to extend the digit set. And if we want to preserve the nice properties of signed representations in conjunction with the ECC, the natural extension of the digit set for the Shamir method is $\mathcal{D} = \{0, \pm1, \pm3\}$, which requires the precomputation of 10 points. To store those points, 400 bytes of memory are required.

Known Methods Using Ten Precomputed Points. The first known method using 10 precomputed points is an extension of the algorithm to create the JSF, called JSF$_3$ which was proposed by Kuang, Zhu and Zhang [KZZ04]. They use the digit set $\mathcal{D} = \{0, \pm1, \pm3\}$ and the resulting AJHD is $121/326 \approx 0.3712$. This methods requires the precomputation of $\{3P, 3Q, P + Q, P - Q, P + 3Q, P - 3Q, 3P + Q, 3P - Q, 3P + 3Q, 3P - 3Q\}$. Another method was proposed by Avanzi [Ava02]. He lets a width-2 window method slide from LtR over the JSF of two scalars to increase the number of zero columns. His method uses the digit set $\mathcal{D} = \{0, \pm1, \pm2, \pm3\}$, but because of the properties of the JSF only the points $\{P + Q, P - Q, P + 2Q, P - 2Q, 2P + Q, 2P - Q, 2P + 3Q, 2P - 3Q, 3P + 2Q, 3P - 2Q\}$ have to be precomputed. The resulting AJHD of this method is $3/8 = 0.3750$. Since both methods reduce the JHW, they are suitable for the Shamir method. However, since they both originate in the JSF, they can only be generated from RtL.

If we want to use even more memory, i.e. extend the digit set even more, the logical choice is the digit set $\mathcal{D} = \{0, \pm1, \pm3, \pm5, \pm7\}$. Now it becomes necessary precompute 38 points, which require 1520 bytes of memory. While this might fit on a smart card, there is another point of concern. If we consider the customary 160-bit scalars, the computational effort (ECDBL and ECADD operations) to precompute those 38 points would be almost as high as the expected effort to compute the actual multi-scalar multiplication. Therefore it is unwise to use larger digit sets and we can conclude that the digit set $\mathcal{D} = \{0, \pm1, \pm3\}$, i.e. the use of ten precomputed points is optimal for the Shamir method.

Now we consider two improvements of the Interleave method which also use 10 precomputed points. The first is to use the wMOF with different widths, namely $w = 4$ for the first scalar and $w = 5$ for the second $((4, 5)$MOF$)$. The used digit sets are $\mathcal{D}_1 = \{0, \pm1, \ldots, \pm7\}$ and $\mathcal{D}_2 = \{0, \pm1, \ldots, \pm15\}$ and the resulting AHDs are $1/5 = 0.2$ and $1/6 \approx 0.1667$ for the first and second scalar, respectively. Then, the average density of ECADD operations is $11/30 \approx 0.3666$. The points to precompute are $\{3P, 3Q, 5P, 5Q, 7P, 7Q, 9Q, 11Q, 13Q, 15Q\}$. The second method is to apply a fractional sliding window method on MOF from LtR [SST04, Möl02, Möl04]. The resulting representation uses the degenerated digit set $\mathcal{D} = \{0, \pm1, \ldots, \pm2^{w-1} + m\}$ and the AHD is $1/(w + \frac{m+1}{2^{w-1}} + 1)$. In order to obtain 10 precomputed points we chose $w = 4$ and $m = 3$ for both scalars. The resulting digit set is $\mathcal{D} = \{0, \pm1, \ldots, \pm11\}$ and the AJHD is $2/11 \approx 0.1818$ for each scalar. This leads to an average density of ECADD operations of $4/11 \approx 0.3636$. Also we have to precompute the points $\{3P, 3Q, 5P, 5Q, 7P, 7Q, 9P, 9Q, 11P, 11Q\}$.

From this one can see that in the case of ten precomputed points the Interleave method currently wins over the Shamir method.

4   Proposed Scheme

In this section we describe the proposed scheme. At first glance our scheme is similar to [HKPR04], namely the main idea is to apply a sliding window method (SWM) with different widths on the MOF of both scalars from LtR. The difference is that we chose a larger digit set and can therefore use larger window widths. For the reasons explained in Section 3 we chose the digit set $\mathcal{D} = \{0, \pm1, \pm3\}$ and therefore need ten precomputed points. The algorithm is divided in three parts: the *Main Routine*, the *Calculation of Z* and the *Conversion Routine*. Further, the recoding can be performed with the knowledge of at most 6 bits of each scalar at a time and we will show that the resulting AJHD is $239/661$ with the method of stochastic processes.

4.1   First Considerations

First, we want to examine how we can use the MOF representation to decrease the JHW. The first MOF property implies that the absolute value of any w consecutive MOF bits is at most $2^{w-1}-1$. Therefore, if we take any w consecutive MOF bits it is possible to represent them using $w-1$ zero entries and 1 non-zero entry with absolute value of at most $2^{w-1}-1$. Since we want to use the digit set $\mathcal{D} = \{0, \pm 1, \pm 3\}$, $w = 3$ holds in our case and by extending this to two scalars, we get

Lemma 1. *Given two MOF representations, a SWM can create at most two consecutive zero columns without exceeding the digit set $\mathcal{D} = \{0, \pm 1, \pm 3\}$. After that, at least one non-zero column must follow.*

Next, we are interested in the position of the columns which are candidates to become zero.

Lemma 2. *Let μ_0 and μ_1 be two k-bit MOF representations. Further, let f_0 and f_1 be the digit of the least non-zero entry of μ_0 and μ_1 respectively. The set*

$$Z := \{k-1, \ldots, 0\} \setminus \{f_0, f_1\}$$

contains the indices of the columns which are candidates to become zero columns.

Note that for two scalars, we have to scan at least three and at most four columns to create two zero columns.

4.2   The Main Routine

The purpose of this part is to decide on the window width used in a certain step. The widths and the required number of zero columns to create are chosen such that the resulting JHD of the recoded window increases from step to step. In other words, at first we try a width which results in a low JHD and if that fails, we increase the width and accept slightly worse JHD. Table 1 shows the sequence in which the widths and the required zero columns are chosen.

If a recoding with one of the first three widths is possible we recode the window, write it out and proceed to the next column. Otherwise after using the last width, where a recoding is always possible, we check the following two conditions to decide how to proceed.

1. If the last two columns remained unchanged after the recoding we write out the first two columns and proceed the scan with the third column.
2. If the last column has been changed, but does not contain any entries equal to ± 3 we write out the first three columns and proceed with the last column.

If those two conditions fail we write out all four columns and proceed with the next column.

However, in the case where we reuse an already recoded column, some problems might occur. Now, it is no longer guaranteed that adjacent non-zero bits have opposite signs. Therefore, Lemma 1 doesn't hold anymore and we have to reduce it to

Lemma 3. *If we reuse a converted column, a SWM can create at most one consecutive zero column without exceeding the digit set $\mathcal{D} = \{0, \pm 1, \pm 3\}$. After that at least one non-zero column must follow.*

According to Lemma 2 now we have to scan at least two and at most three columns in order to create one zero column. Therefore we use a different sequence of widths as shown in Table 1.

Table 1. Sequence of window widths with and without reusing

	without reusing				with reusing			
Sequence of conversion	1.	2.	3.	4.	1.	2.	3.	4.
zero columns required	1	2	3	2	1	1	2	1
window width	1	3	5	4	1	2	4	3
resulting JHD	0	0.33	0.4	0.5	0	0.5	0.5	0.66

Again, if a recoding using one of the first three widths is possible we recode the window, write it out and proceed to the next column. Otherwise we apply the fourth conversion and perform the same checks as above. Note that in all cases where we don't reuse an already converted column, Lemma 1 holds again in the next step.

4.3   The Calculation of Z

This method computes the number of zero columns which can be created in a certain window. Therefore it is used by the main routing to decide whether a certain width should be used or not. Further it computes the positions of the columns to become zero, which are needed by the conversion routine. Hence, at first we calculate the set Z according to Lemma 2. Next, we select a set $\tilde{Z} \subset Z$ which represents the columns that will actually be converted to zero. This choice is performed according to Lemma 1 or Lemma 3. If we have more than one possibility for $\tilde{Z}$, we start picking the leftmost candidates first. In the following examples let $\bar{x} = -x$.

Example 1.

a) *Without reusing. Let $\mu_0 = \bar{1}01\bar{1}1$, $\mu_1 = 10\bar{1}00$. Therefore $f_0 = 0$ and $f_1 = 2$ holds.*

$$\text{Lemma 2} \Longrightarrow Z := \{4, 3, 2, 1, 0\} \setminus \{2, 0\} = \{4, 3, 1\} \overset{\text{Lemma 1}}{\Longrightarrow} \tilde{Z} = \{4, 3, 1\}$$

b) *With reusing. Let $\mu_0 = \bar{1}\bar{1}\bar{1}1$, $\mu_1 = \bar{1}0\bar{1}1$. Therefore $f_0 = 0$ and $f_1 = 0$ holds.*

$$\text{Lemma 2} \Longrightarrow Z := \{3, 2, 1, 0\} \setminus \{0\} = \{3, 2, 1\} \overset{\text{Lemma 3}}{\Longrightarrow} \tilde{Z} = \{3, 1\}$$

4.4   The Conversion Routine

This part performs the actual recoding of the window. At this point we know which columns shall become zero, therefore it is possible to recode each scalar separately. Each window is scanned from LtR and if a non-zero entry which should become zero is detected, we scan for the next non-zero entry on the right and apply one of the following conversions.

$$\text{(1)}\quad 100\ldots 0\bar{1} \mapsto 011\ldots 11 \qquad\qquad \text{(2)}\quad \bar{1}00\ldots 01 \mapsto 0\bar{1}\bar{1}\ldots \bar{1}\bar{1}$$

$$\text{(3)}\quad 100\ldots 01 \mapsto 03\bar{1}\ldots \bar{1}\bar{1} \qquad\qquad \text{(4)}\quad \bar{1}00\ldots 0\bar{1} \mapsto 0\bar{3}1\ldots 11$$

Note, that because of Lemma 2 we are always able to find a non-zero entry to the right in the current window.

Example 2.

a) Let $\mu_0 = \bar{1}01\bar{1}1$, $\mu_1 = 10\bar{1}00$, $\tilde{Z} = \{4, 3, 1\}$. Applying (1) − (4) yields

$$\bar{1}01\bar{1}1 \xrightarrow{(2)} 0\bar{1}\bar{1}\bar{1}1 \xrightarrow{(4)} 00\bar{3}\bar{1}1 \xrightarrow{(2)} 00\bar{3}0\bar{1}$$
$$\uparrow\qquad\qquad \uparrow\qquad\qquad \uparrow$$
$$10\bar{1}00 \xrightarrow{(1)} 01100 \xrightarrow{(3)} 00300 \longmapsto 00300$$
$$\uparrow\qquad\qquad \uparrow\qquad\qquad \uparrow$$

b) Let $\mu_0 = \bar{1}\bar{1}\bar{1}1$, $\mu_1 = \bar{1}0\bar{1}1$, $\tilde{Z} = \{3, 1\}$. Applying (1) − (4) yields

$$\bar{1}\bar{1}\bar{1}1 \xrightarrow{(4)} 0\bar{3}\bar{1}1 \xrightarrow{(2)} 0\bar{3}0\bar{1} \qquad \bar{1}0\bar{1}1 \xrightarrow{(4)} 0\bar{3}11 \xrightarrow{(3)} 0\bar{3}03$$
$$\uparrow\qquad\qquad \uparrow \qquad\qquad\qquad\qquad \uparrow\qquad\qquad \uparrow$$

4.5 Implementation

The implementation of the three parts of the proposed scheme can be found in Algorithms 1, 2 and 3. They use the following notations: The variables u and l denote the first and the last index of the current window, respectively. The variable c denotes the current case, namely $c = 0$ if we are reusing an already converted column and $c = 1$ otherwise. The set Z to determine which columns should be converted, is represented as a k-bit array z, where $z_j = 1$, if the j-th column in the current window is to be converted and $z_j = 0$, otherwise. Here k is the width of the current window and $j = k - 1, \ldots, 0$. The notation $d_{i,j}$ denotes the j-th bit of the i-th scalar and substrings are denoted by $d_{0,u..l} := (d_{0,u}, d_{0,u-1}, \ldots, d_{0,l})$. Also, $\ominus$ denotes the bitwise subtraction which is used for the on-the-fly MOF generation.

4.6 Average Joint Hamming Density

The next step is to prove, that the representation generated by the proposed scheme indeed results in an AJHD of $239/661 \approx 0.3615$. We will calculate the AJHD using Markov Chains [Häg02]. Figure 1 shows the transition graph of the proposed scheme. Each state indicates the number of columns currently scanned, the number of columns which are reused (the boxed ones) and the probability with which the state changes into another. Whenever a recoding was performed, we jump back to state 1. Those changes are indicated by arrows with a dot at the end.

The transition probabilities are given by the matrix $(p_{ij}) := P(S_i \mapsto S_j)$, where $S_i \mapsto S_j$ indicates that state S_i changes into S_j. Those numbers were

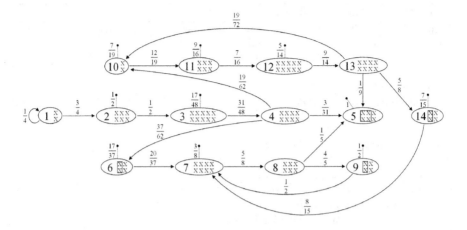

Fig. 1. Transition graph

obtained by checking all cases. Also we need the matrices (t_{ij}) which contains the total number of columns written out by the algorithm if $S_i \mapsto S_j$ and (n_{ij}) which contains the number of non-zero columns written out if $S_i \mapsto S_j$, $i, j = 1, \ldots, 14$. The non-zero entries of those three matrices as well as the line in Algorithm 1 where the changes of states occur are summarized in Table 2.

Since this Markov chain is irreducible and aperiodic, it exists a stationary distribution

$$\pi = \left(\frac{976}{2885}, \frac{732}{2885}, \frac{366}{2885}, \frac{1891}{23080}, \frac{227}{17310}, \frac{2257}{46160}, \frac{323}{8655}, \frac{323}{13848}, \frac{323}{17310}, \frac{76}{2885}, \frac{48}{2885}, \frac{21}{2885}, \frac{27}{5770}, \frac{27}{9232} \right)$$

Table 2. Non-zero entries of the matrices p_{ij}, t_{ij} and n_{ij}

line	$S_i \mapsto S_j$	p_{ij}	t_{ij}	n_{ij}	line	$S_i \mapsto S_j$	p_{ij}	t_{ij}	n_{ij}
1	$S_1 \mapsto S_1$	1/4	1	0	1	$S_3 \mapsto S_1$	17/48	5	2
1	$S_6 \mapsto S_1$	17/37	2	1	1	$S_7 \mapsto S_1$	3/8	4	2
1	$S_9 \mapsto S_1$	1/2	2	1	1	$S_{12} \mapsto S_1$	5/14	5	2
1	$S_{10} \mapsto S_1$	7/19	1	0	1	$S_3 \mapsto S_4$	31/48	0	0
1	$S_{14} \mapsto S_1$	7/15	2	1	1	$S_7 \mapsto S_8$	5/8	0	0
1/1	$S_6 \mapsto S_7$	20/37	0	0	1	$S_{12} \mapsto S_{13}$	9/14	0	0
1/1	$S_9 \mapsto S_7$	1/2	0	0	1	$S_4 \mapsto S_5$	3/31	2	1
1/1	$S_{14} \mapsto S_7$	8/15	0	0	1	$S_8 \mapsto S_5$	1/5	1	1
1	$S_1 \mapsto S_2$	3/4	0	0	1	$S_{13} \mapsto S_5$	1/9	2	1
1	$S_{10} \mapsto S_{11}$	12/19	0	0	1	$S_4 \mapsto S_6$	37/62	3	1
1	$S_2 \mapsto S_1$	1/2	3	1	1	$S_8 \mapsto S_9$	4/5	2	1
1	$S_5 \mapsto S_1$	1	3	1	1	$S_{13} \mapsto S_{14}$	5/8	3	1
1	$S_{11} \mapsto S_1$	9/16	3	1	1	$S_4 \mapsto S_{10}$	19/62	4	2
1	$S_2 \mapsto S_3$	1/2	0	0	1	$S_{13} \mapsto S_{10}$	19/72	4	2
1	$S_{11} \mapsto S_{12}$	7/16	0	0					

Algorithm 1. The Main Routine

Require: two n-bit scalars d_0 and d_1 in their binary representation
Ensure: recoded representation δ_0 and δ_1

1: $d_{0,-1} \hookleftarrow 0$; $d_{1,-1} \hookleftarrow 0$; $d_{0,n} \hookleftarrow 0$; $d_{1,n} \hookleftarrow 0$
2: $u \hookleftarrow n$; $c \hookleftarrow 1$
3: **while** $u > 0$ **do**
4:     **while** $d_{0,u} = d_{0,u-1} \wedge d_{1,u} = d_{1,u-1} \wedge u > 0$ **do**
5:         $\mu_{0,u} \hookleftarrow 0$; $\mu_{1,u} \hookleftarrow 0$
6:         $u \hookleftarrow u - 1$; $c \hookleftarrow 1$
7:     **end while**
8:     $l \hookleftarrow u - 1 - c$
9:     $\mu_{0,u+c-1..l} \hookleftarrow d_{0,u+c-1..l} \ominus d_{0,u+c-2..l-1}$
10:     $\mu_{1,u+c-1..l} \hookleftarrow d_{1,u+c-1..l} \ominus d_{1,u+c-2..l-1}$
11:     $z \hookleftarrow \text{calculateZ}(\mu_{0,u..l}, \mu_{1,u..l}, c)$
12:     **if** $z_{u-l} + \ldots + z_0 \geq 1 + c \vee l \leq 0$ **then**
13:         $(\mu_{0,u...l}, \mu_{1,u...l}) \hookleftarrow \text{convert}(\mu_{0,u..l}, \mu_{1,u..l}, z)$
14:         $u \hookleftarrow u - 2 - c$; $c \hookleftarrow 1$
15:     **else**
16:         $l \hookleftarrow u - 3 - c$
17:         $\mu_{0,u+c-1..l} \hookleftarrow d_{0,u+c-1..l} \ominus d_{0,u+c-2..l-1}$
18:         $\mu_{1,u+c-1..l} \hookleftarrow d_{1,u+c-1..l} \ominus d_{1,u+c-2..l-1}$
19:         $z \hookleftarrow \text{calculateZ}(\mu_{0,u..l}, \mu_{1,u..l}, c)$
20:         **if** $z_3 = 1 \wedge z_2 = 0 \wedge z_1 = 1 \wedge z_0 = 0$ **then**
21:             $(\mu_{0,u...l}, \mu_{1,u...l}) \hookleftarrow \text{convert}(\mu_{0,u..l}, \mu_{1,u..l}, z)$
22:             $u \hookleftarrow u - 4 - c$; $c \hookleftarrow 1$
23:         **else**
24:             $l \hookleftarrow u - 2 - c$
25:             $\mu_{0,u+c-1..l} \hookleftarrow d_{0,u+c-1..l} \ominus d_{0,u+c-2..l-1}$
26:             $\mu_{1,u+c-1..l} \hookleftarrow d_{1,u+c-1..l} \ominus d_{1,u+c-2..l-1}$
27:             $z \hookleftarrow \text{calculateZ}(\mu_{0,u..l}, \mu_{1,u..l}, c)$
28:             $(\mu_{0,u..l}, \mu_{1,u..l}) \hookleftarrow \text{convert}(\mu_{0,u..l}, \mu_{1,u..l}, z)$
29:             **if** $\mu_{0,l+1..l} = d_{0,l+1..l} \ominus d_{0,l..l-1} \wedge \mu_{1,l+1..l} = d_{1,l+1..l} \ominus d_{1,l..l-1}$ **then**
30:                 $u \hookleftarrow u - 1 - c$; $c \hookleftarrow 1$
31:             **else if** $\mu_{0,l} \neq \pm 3 \wedge \mu_{1,l} \neq \pm 3 \wedge (\mu_{0,l}, \mu_{1,l}) \neq (0,0)$ **then**
32:                 $u \hookleftarrow u - 2 - c$; $c \hookleftarrow 0$
33:             **else**
34:                 $u \hookleftarrow u - 3 - c$; $c \hookleftarrow 1$
35:             **end if**
36:         **end if**
37:     **end if**
38: **end while**
39: **return** μ_0, μ_1.

Using the stationary distribution π and the matrices (p_{ij}), (t_{ij}) and (n_{ij}) we can calculate the AJHD as follows. According to the definition of the AJHD, we need the average number of (non-zero) columns written out by the algorithm for any possible transition $S_i \mapsto S_j$, $i, j = 1, \ldots 14$. For one fixed transition $S_{\tilde{i}} \mapsto S_{\tilde{j}}$, these numbers obviously are $(n_{\tilde{i}\tilde{j}} \cdot p_{\tilde{i}\tilde{j}})$ $t_{\tilde{i}\tilde{j}} \cdot p_{\tilde{i}\tilde{j}}$. If we consider a whole state $S_{\tilde{i}}$,

---

Algorithm 2. Calculation of z *calculateZ*

Require: two k-bit MOF strings μ_0 and μ_1 and the current case c
Ensure: the vector z
1: **for** $i = 0$ to 1 **do**
2:     $f_i \leftarrow -1$
3:     **for** $j = k - 1$ down to 0 **do**
4:       **if** $\mu_{i,j} \neq 0$ **then**
5:         $f_i \leftarrow j$
6:       **end if**
7:     **end for**
8: **end for**
9: $r \leftarrow 0$
10: **for** $j = k - 1$ down to 0 **do**
11:     **if** $j = f_0 \vee j = f_1 \vee r = 2$ **then**
12:       $z_j \leftarrow 0; r \leftarrow 0$
13:     **else**
14:       $z_j \leftarrow 1; r \leftarrow r + 1$
15:     **end if**
16:     **if** $c = 0 \wedge j = k - 1 \wedge z_{k-1} = 1$ **then**
17:       $r \leftarrow 2$
18:     **end if**
19: **end for**
20: **return** z.

Algorithm 3. Conversion routine *convert*

Require: two k-bit MOF strings μ_0 and μ_1 and the columns to convert z
Ensure: recoded representation of μ_0 and μ_1
1: **for** $i = 0$ to 1 **do**
2:     **for** $j = k - 1$ down to 0 **do**
3:       **if** $z_j = 1 \wedge \mu_{i,j} \neq 0$ **then**
4:         $s \leftarrow j - 1$
5:         **while** $\mu_{i,s} = 0$ **do**
6:           $s \leftarrow s - 1$
7:         **end while**
8:         **if** $\mu_{i,j} = -\mu_{i,s}$ **then**
9:           **for** $t = j - 1$ down to s **do**
10:             $\mu_{i,t} \leftarrow \mu_{i,j}$
11:           **end for**
12:           $\mu_{i,j} \leftarrow 0$
13:         **else if** $\mu_{i,j} = \mu_{i,s}$ **then**
14:           **for** $t = j - 2$ down to s **do**
15:             $\mu_{i,t} \leftarrow -\mu_{i,j}$
16:           **end for**
17:           $\mu_{i,j-1} \leftarrow 3 \cdot \mu_{i,j}; \mu_{i,j} \leftarrow 0$
18:         **end if**
19:       **end if**
20:     **end for**
21: **end for**
22: **return** μ_0, μ_1.

---

we have to add all values $(n_{\tilde{i}j} \cdot p_{\tilde{i}j}) \, t_{\tilde{i}j} \cdot p_{\tilde{i}j}$, $j = 1, \ldots, 14$. Finally if we consider all states S_i, $i = 1, \ldots, 14$ we have to multiply this sum with π_i, the probability that the algorithm currently is in this state and add them together. The AJHD then is the quotient of the value for the non-zero columns and the value for all columns, namely

$$\text{AJHD} = \frac{\sum\limits_{i=1}^{14} \pi_i \sum\limits_{j=1}^{14} n_{ij} \cdot p_{ij}}{\sum\limits_{i=1}^{14} \pi_i \sum\limits_{j=1}^{14} t_{ij} \cdot p_{ij}} = \frac{239}{661} \approx 0.3615733$$

Because such calculations involve a great number of difficult to estimate values, it is very likely that some error occurs. However, we are happy to report that this AJHD was confirmed by experimental results. While for 160-bit scalars the estimated AJDH was 0.3636836, for a larger bit length it converged against the calculated value.

5   Comparison

In this section we want to compare the average number of ECADD operations required for computing $uP + vQ$ of the proposed scheme and the methods of Section 3. Table 3 shows these values and also the direction in which the scalar are recoded, i.e. LtR od RtL.

According to Table 3 the proposed scheme requires the least average number of additions and is therefore the first scheme with which the Shamir method wins over the Interleave method. Compared to the first two methods [Ava02, KZZ04] the memory usage for the recoding is reduced due to the LtR generation. The problem with the last two methods is that the underlying representations have been proven to be minimal [Ava04, Möl04]. Therefore it is not possible to further reduce the average number of additions using methods which reduce the AHD, while for methods which reduce the AJHD a minimal representation is still unknown.

Table 3. Average number of additions and direction of recoding

Scheme	avg. number of ECADD	direction
Shamir+[Ava02]	$\frac{3}{8}n = 0.3750n$	RtL
Shamir+[KZZ04]	$\frac{121}{326}n \approx 0.3712n$	RtL
Interleave+(4,5)MOF	$(\frac{1}{6} + \frac{1}{5})n \approx 0.3666n$	LtR
Interleave+[Möl04]	$(\frac{2}{11} + \frac{2}{11})n \approx 0.3636n$	LtR
Shamir+Section 4	$\frac{239}{661}n \approx 0.3615n$	LtR

6   Conclusion

In this paper we proposed a new algorithm to speed up the calculation of $uP+vQ$ using the Shamir method. The main point was extending the digit set of the JSF to $\mathcal{D} = \{0, \pm 1, \pm 3\}$. We proved that the AJHD of our scheme is $239/661 \approx$ 0.3615, which is superior to any known method which uses ten precomputed points. The proposed scheme is the first to enhance the Shamir method such that it wins over the Interleave method and compared to the JSF, the multi-scalar multiplication can be sped up by more than 10%. Due to the LtR fashion of our algorithm, the memory consumption for the recoding is reduced and we need only the knowledge of 6 joint bits of the binary representations to generate the new representation. Future work may include an improvement of the AJHD and a generalisation to an arbitrary number of scalars.

References

[Ava02]   Avanzi, R., *On multi-exponentiation in cryptography*, Cryptology ePrint Archive: Report 2002/154, 2002, available at
            http://eprint.iacr.org/2002/154/

[Ava04]   Avanzi, R., *A Note on the Signed Sliding Window Integer Recoding and a Left-to-Right Analogue*, Selected Areas in Cryptography - SAC 2004, LNCS 3357, pp. 130-143

[BSS99]   Blake, I., Seroussi, G., and Smart, N., *Elliptic Curves in Cryptography*, Cambridge University Press, 1999.

[CMO98]   Cohen, H., Miyaji, A., Ono, T., *Efficient Elliptic Curve Exponentiation Using Mixed Coordinates*, Advances in Cryptology - ASIACRYPT '98, LNCS1514, (1998), 51-65.

[ElG85]   ElGamal, T., *A Public Key Cryptosystem and a Signature Scheme Based on Discrete Logarithms*, IEEE Transactions on Information Theory, Vol. 31, IEEE 1985, pp. 469-472.

[GHPT03]   Grabner, P., Heuberger, C., Prodinger, H., Thuswaldner J., *Analysis of linear combination algorithms in cryptography*, available at
            http://www.opt.math.tu-graz.ac.at/~cheub/publications/

[Gor98]   Gordon, D., *A survey of fast exponentiation methods*, Journal of Algorithms, vol.27, (1998), 129-146.

[Häg02]   Häggström, O., *Finite Markov Chains and Algorithmic Applications*, London Mathematical Society Student Texts 52, Cambridge University Press, (2002).

[HKPR04]   Heuberger, C., Katti, R., Prodinger, H., Ruan, X., *The Alternating Greedy Expansion and Applications to Left-To-Right Algorithms in Cryptography* available at
            http://www.opt.math.tu-graz.ac.at/~cheub/publications/

[Kob87]   Koblitz, N., *Elliptic Curve Cryptosystems*, Math. Comp. 48, (1987), 203-209.

[KZZ04]   Kuang, B., Zhu, Y., Zhang, Y., *An Improved Algorithm for uP+vQ using JSF_3*, Applied Cryptography and Network Security - ACNS 2005, LNCS 3089, pp. 467-478

[Mil86]   Miller, V.S., *Use of Elliptic Curves in Cryptography*, Advances in Cryptology - CRYPTO '85, LNCS218, (1986), 417-426.

[MOC97]   Miyaji, A., Ono, T., and Cohen, H., *Efficient Elliptic Curve Exponentiation*, Information and Communication Security, ICICS 1997, LNCS 1334, (1997), 282-291.

[MO90]   Morain, F., Olivos, J., *Speeding Up the Computations on an Elliptic Curve using Addition-Subtraction Chains*, Informa. Theor. Appl., 24, (1990), pp.531-543.

[Möl01]   Möller, B., *Algorithms for Multi-exponentiation*, Selected Areas in Cryptography - SAC 2001, LNCS 2259, pp. 165-180

[Möl02]   Möller, B., *Improved Techniques for Fast Exponentiation*, Information Security and Cryptology ICISC 2002. LNCS 2587, pp. 298312

[Möl04]   Möller, B., *Fractional Windows Revisited: Improved Signed-Digit Representations for Efficient Exponentiation*, Information Security and Cryptology ICISC 2004, to appear.

[OSST04]   Okeya, K., Schmidt-Samoa, K., Spahn, C., Takagi, T., *Signed Binary Representations Revisited*, Advances in Cryptology - CRYPTO 2004, LNCS 3152, pp.123-139, available at
            http://eprint.iacr.org/2004/195/

[Pro03]   Proos, J., *Joint Sparse Forms and Generating Zero Columns when Combing*, Technical Report of the Centre for Applied Cryptographic Research, University of Waterloo - CACR, CORR 2003-23, 2003, available at
            http://www.cacr.math.uwaterloo.ca.

[SST04]   Schmidt-Samoa, K., Semay, O., Takagi, T., *Analysis of Some Efficient Window Methods and their Application to Elliptic Curve Cryptosystems*, Technical Report No. TI-3/04, 16. August 2004.

[Sol00]   Solinas, J.A., *Efficient Arithmetic on Koblitz Curves*, Design, Codes and Cryptography, 19, (2000), 195-249.

[Sol01]   Solinas, J.A., *Low-weight binary representations for pairs of integers*, Technical Report of the Centre for Applied Cryptographic Research, University of Waterloo - CACR, CORR 2001-41, 2001, available at
            http://www.cacr.math.uwaterloo.ca

Side Channel Attacks on Message Authentication Codes

Katsuyuki Okeya[1] and Tetsu Iwata[2]

[1] Hitachi, Ltd., Systems Development Laboratory,
1099, Ohzenji, Asao-ku, Kawasaki, 215-0013, Japan
ka-okeya@sdl.hitachi.co.jp
[2] Dept. of Computer and Information Sciences,
Ibaraki University, 4–12–1 Nakanarusawa,
Hitachi, Ibaraki 316-8511, Japan
iwata@cis.ibaraki.ac.jp

Abstract. Side channel attacks are a serious menace to embedded devices with cryptographic applications which are utilized in sensor and ad hoc networks. In this paper we show that side channel attacks can be applied to message authentication codes, even if the countermeasure is applied to the underlying block cipher. In particular, we show that EMAC, OMAC, and PMAC are vulnerable to our attack. Based on simple power analysis, we show that several key bits can be extracted, and based on differential power analysis, we present selective forgery against these MACs. Our results suggest that protecting block ciphers against side channel attacks is not sufficient, and countermeasures are needed for MACs as well.

Keywords: Side Channel Attacks, MACs, Selective Forgery, SPA, DPA.

1 Introduction

Ubiquitous computing devices are penetrating in our daily life. A sensor network is such an example. While some sensors have batteries, others are supplied with electricity from outside. However, side channel attacks (SCA) are a serious menace to such embedded devices with cryptographic applications. In SCA the attacker reveals secret information using side channel information such as power consumption while the victim device performs cryptographic applications [Koc96, KJJ99].

An ad hoc network is also useful for mobile devices to communicate each other. In many environments including the ad hoc network, authentication and integrity check mechanism are important to prevent against impersonation and substitution/alternation of messages. Since such devices are equipped with scarce computational resources only, authentication based on the symmetric ciphers, namely message authentication code (MAC) is often utilized.

Nowadays some IC chips are equipped with tamper-resistant modules as countermeasures against SCA, especially countermeasures on a block cipher are utilized. Under this situation, MACs are implemented as a software program using

R. Molva, G. Tsudik, and D. Westhoff (Eds.): ESAS 2005, LNCS 3813, pp. 205–217, 2005.
© Springer-Verlag Berlin Heidelberg 2005

the tamper-resistant block cipher module. Whereas many researches have been made on SCA against block ciphers, little attention has been given to SCA against MACs so far.

This paper discusses the security of MACs with tamper-resistant block cipher.

1.1   Previous Works

First, Vaudenay demonstrated a side-channel attack against CBC encryption mode with CBC-PAD [Vau02]. Given an oracle which reveals whether or not the plaintext (corresponding to some altered ciphertext) is correctly padded (side channel information), the paper [Vau02] showed that one can efficiently recover the plaintext. This attack is often referred to as the padding oracle attack. Later, Black and Urtubia described an improvement to the Vaudenay's attack [BU02]. The paper [BU02] generalized the attack to other encryption schemes showing that other common methods for symmetric encryption (CTR, OFB, CFB, and stream ciphers) all possess the required weaknesses which permit this type of attack. Paterson and Yau considered CBC mode of ISO/IEC 10116 [PY04]. Klima and Rosa described how the CBC mode in the PKCS#7 can be attacked [KR03].

Möler [Möl04] pointed out problems with the CBC-based ciphersuites in SSL 3.0 and TLS 1.0. In fact, TLS 1.0 has different error codes for incorrect MACs ('decryption_failed' and 'bad_record_mac'), the attacker can utilize it as the padding oracle. Because of the menace of the padding oracle attack, these error codes were unified. Next, Canvel, Hiltgen, Vaudenay, and Vaugnoux showed that the attack is actually applicable against the popular implementations of SSL/TLS for password interception [CHVV03].

1.2   Contribution of This Paper

The lesson from the above results is that a strong message integrity check is needed, and padding schemes or obscure message encoding does not seem to help in preventing these kind of attacks.

Message integrity checks are often realized by using a Massage Authentication Code, or a MAC for short. A MAC takes a secret key and a message to produce a fixed length output, called a tag. This tag is then used to check if the message is altered during the transmission, or storage.

There are several well known MACs based on block ciphers, for example, we have CBC MAC [FIPS94, ISO99], EMAC [BBB+95], OMAC [IK03], and PMAC [BR02].

CBC MAC is a widely used standard MAC, and EMAC was developed for the RACE project. OMAC is recommended by NIST under the name of CMAC [NIST], and also is in the process of approval at IEEE 802.16 Task Group e (Mobile WirelessMAN) [IEEE].

In this paper, we show that side channel attacks can be applied to these MACs, even if the countermeasure is applied to the underlying block cipher. In particular, we show that EMAC, OMAC, and PMAC are vulnerable to our attack. As SCAs, we consider Simple Power Analysis (SPA) and Differential Power Analysis (DPA) [Koc96, KJJ99] against these MACs. Based on SPA, we show that two bits of information (which is equivalent to two bits of the key) is extracted in OMAC, and $\log m$ bits of information (which is equivalent to $\log m$ bits of the key) can be extracted in PMAC, where m denotes the maximum length of the messages that the attacker obtains. For DPA, we present selective forgery against EMAC, OMAC, and PMAC, assuming that the attacker is in a chosen plaintext attack scenario. We note that this attack is much stronger and useful than the standard existential forgery, since the attacker can choose the message. For example the message may be a contract that is beneficial to the attacker. Furthermore, we allow the attacker to choose the target message *before* the key is chosen.

Our results suggest that protecting block ciphers against side channel attacks is not sufficient, and the countermeasure is needed for implementations of MACs as well.

2   Block Ciphers and Message Authentication Codes

Notation. If x is a string then $|x|$ denotes its length in bits. If x and y are two equal-length strings, then $x \oplus y$ denotes the XOR of x and y. If x and y are strings, then $x \circ y$, or simply xy, denotes their concatenation. Let $x \leftarrow y$ denote the assignment of y to x. If X is a set, let $x \overset{R}{\leftarrow} X$ denote the process of uniformly selecting at random an element from X and assigning it to x. For positive number n, $\{0,1\}^n$ is the set of all binary strings of length n. We also denote $(\{0,1\}^n)^+$ the set all binary strings of length positive multiple of n, and $\{0,1\}^*$ the set all binary strings.

Block Ciphers. A block cipher $E : \{0,1\}^k \times \{0,1\}^n \rightarrow \{0,1\}^n$ is a function from k-bit keys and n-bit blocks to n-bit blocks. We use the notation $E_K(X)$ as shorthand for $E(K,X)$. A block cipher is a family of permutations; that is, for each key $K \in \{0,1\}^k$, $E_K(\cdot)$ is a permutation on $\{0,1\}^n$. We call k the key length of E and we call n the block length.

MACs and Their Security. A message authentication code or MAC $\mathcal{MA} = (\mathcal{K}, \mathcal{T}, \mathcal{V})$ consists of three algorithms and is defined for some message space MSG and some tag length τ. The randomized key generation algorithm $\mathcal{K}$ takes no input and returns a random key K. The stateless and deterministic tagging algorithm takes a key K and a message $M \in$ MSG as input and returns a tag $T \in \{0,1\}^\tau$; we write $\mathcal{T}_K(M) = T$. The stateless and deterministic verification algorithm takes a key K, a message $M \in$ MSG, and a candidate tag $T \in \{0,1\}^\tau$ as input and returns a bit b; we write $\mathcal{V}_K(M,T) = b$. For consistency, we require that for all keys K and messages M, $\mathcal{V}_K(M, \mathcal{T}_K(M)) = 1$.

In this paper, we assume that the adversary is in a chosen plaintext attack scenario. Such attacks are possible, for example, if a device such as smart card containing a secret key is physically accessible by the attacker. In this scenario the device acts as oracles, and the attacker A has two oracles: a tagging oracle

$\mathcal{T}_K(\cdot)$ and a verification oracle $\mathcal{V}_K(\cdot, \cdot)$. We say that A *forges* if A makes a query (M^*, T^*) to $\mathcal{V}_K(\cdot, \cdot)$ such that $\mathcal{V}_K(M^*, T^*) = 1$ and A did not make a query M^* to $\mathcal{T}_K(\cdot)$ that resulted in a response T^*. Then

$$\mathbf{Adv}_{\mathcal{MA}}^{\mathrm{mac}}(A) \stackrel{\mathrm{def}}{=} \Pr(K \stackrel{R}{\leftarrow} \mathcal{K} : A^{\mathcal{T}_K(\cdot), \mathcal{V}_K(\cdot, \cdot)} \text{ forges})$$

is defined as the *MAC-advantage* of A against $\mathcal{MA}$. Intuitively, the attacker forges if the integrity of the message is broken, that is, if the adversary is able to obtain the correct (message, tag) pair, where this pair has not previously produced by the oracle.

(M^*, T^*) is called a forgery attempt, and it is called a forgery if $\mathcal{V}_K(M^*, T^*) = 1$. If the attacker can choose M^* at will, then (M^*, T^*) is called a *selective forgery*, otherwise it is called an *existential forgery*.

3   MACs: CBC MAC, EMAC, OMAC and PMAC

CBC MAC [FIPS94, ISO99]. Let $E : \{0,1\}^k \times \{0,1\}^n \to \{0,1\}^n$ be a block cipher. The function CBC is parameterized by E. CBC $: \{0,1\}^k \times (\{0,1\}^n)^+ \to \{0,1\}^n$ takes a key $K \in \{0,1\}^k$ and a message $M \in (\{0,1\}^n)^+$ as input and returns an n-bit string. The algorithm of CBC is described in Figure 1.

EMAC [BBB+95]. Let $E : \{0,1\}^k \times \{0,1\}^n \to \{0,1\}^n$ be a block cipher and let $\tau \leq n$ be the tag length. We write EMAC$[E, \tau]$ if we use E and τ as parameters. The EMAC$[E, \tau]$ key generation algorithm EMAC-$\mathcal{K}$ returns two independent random k-bit keys K_1 and K_2. The EMAC$[E, \tau]$ tagging algorithm EMAC-$\mathcal{T}$ takes K_1 and K_2 and a message $M \in (\{0,1\}^n)^+$ as input and returns a τ-bit tag T. The algorithm of EMAC-$\mathcal{T}$ is described in Figure 2.

The EMAC$[E, \tau]$ verification algorithm EMAC-$\mathcal{V}$ takes K_1 and K_2, a message $M \in (\{0,1\}^n)^+$, and a tag $T \in \{0,1\}^\tau$ as input and returns 1 iff EMAC-$\mathcal{T}_{K_1, K_2}(M) = T$.

OMAC [IK03]. Let $E : \{0,1\}^k \times \{0,1\}^n \to \{0,1\}^n$ be a block cipher and let $\tau \leq n$ be the tag length. We write OMAC$[E, \tau]$ if we use E and τ as parameters. The OMAC$[E, \tau]$ key generation algorithm OMAC-$\mathcal{K}$ returns a random k-bit key K. The OMAC$[E, \tau]$ tagging algorithm OMAC-$\mathcal{T}$ takes a k-bit key K and

---

Algorithm CBC$_K(M)$
Let $M[1] \cdots M[m] \leftarrow M$ where $|M[i]| = n$
$Y[0] \leftarrow 0^n$
for $i \leftarrow 1$ **to** m **do**
      $Y[i] \leftarrow E_K(M[i] \oplus Y[i-1])$
return $Y[m]$

Fig. 1. CBC$_K(M)$

```
Algorithm EMAC-T_{K_1,K_2}(M)
Y[m] ← CBC_{K_1}(M)
T ← the leftmost τ bits of E_{K_2}(Y[m])
return T
```

Fig. 2. EMAC-$\mathcal{T}_{K_1,K_2}(M)$

```
Algorithm OMAC-T_K(M)
L ← E_K(0^n)
Let M = M[1] ⋯ M[m], where |M[i]| = n for 1 ≤ i ≤ m − 1
Y[m − 1] ← CBC_K(M[1], ... , M[m − 1])
if |M[m]| = n then X[m] ← Y[m − 1] ⊕ M[m] ⊕ L · u
              else X[m] ← Y[m − 1] ⊕ pad(M[m]) ⊕ L · u^2
T ← E_K(X[m])
return the leftmost τ bits of T
```

Fig. 3. OMAC-$\mathcal{T}_K(M)$

a message $M \in \{0,1\}^*$ as input and returns a τ-bit tag T. The algorithm of OMAC-$\mathcal{T}$ is described in Figure 3. In Figure 3, the padding function $\mathsf{pad}(\cdot)$ takes $M \in \{0,1\}^*$ such that $|M| \leq n$ and is defined as follows:

$$\mathsf{pad}(M) = M \circ 10^{n-1-(|M| \bmod n)} \tag{1}$$

$L \cdot \mathsf{u}$ is defined as follows:

$$L \cdot \mathsf{u} = \begin{cases} L \ll 1 & \text{if } \mathsf{msb}(L) = 0, \\ (L \ll 1) \oplus \mathsf{Cst}_n & \text{if } \mathsf{msb}(L) = 1, \end{cases} \tag{2}$$

where: $\mathsf{msb}(L)$ denotes the most significant bit of L (meaning the left most bit), $L \ll 1$ denotes the left shift of L by one bit (the most significant bit disappears and a zero comes into the least significant bit), and Cst_n is an n-bit constant. For example, $\mathsf{Cst}_{64} = 0^{59}11011$ and $\mathsf{Cst}_{128} = 0^{120}10000111$. $L \cdot \mathsf{u}^2$ is simply $(L \cdot \mathsf{u}) \cdot \mathsf{u}$.

We also note that multiplication by u^{-1} is easy. For any $L \in \{0,1\}^n$, if $\mathsf{lsb}(L) = 0$ then $L \cdot \mathsf{u}^{-1}$ is $L \gg 1$, and if $\mathsf{lsb}(L) = 1$ then $L \cdot \mathsf{u}^{-1}$ is $(L \gg 1) \oplus \mathsf{Cst}'_n$, where: $\mathsf{lsb}(L)$ denotes the least significant bit of L (meaning the right most bit), $L \gg 1$ denotes the right shift of L by one bit (the least significant bit disappears and a zero comes into the most significant bit), and Cst'_n is an n-bit constant. For example, $\mathsf{Cst}'_{128} = 10^{120}1000011$.

The OMAC$[E, \tau]$ verification algorithm OMAC-$\mathcal{V}$ takes a k-bit key K, a message $M \in \{0,1\}^*$, and a tag $T \in \{0,1\}^\tau$ as input and returns 1 iff OMAC-$\mathcal{T}_K(M) = T$.

PMAC [BR02]. Let $E : \{0,1\}^k \times \{0,1\}^n \to \{0,1\}^n$ be a block cipher and let $\tau \leq n$ be the tag length. We write PMAC$[E, \tau]$ if we use E and τ as parameters. The PMAC$[E, \tau]$ key generation algorithm PMAC-$\mathcal{K}$ returns a random k-bit key K.

Algorithm PMAC-$\mathcal{T}_K(M)$

$L \leftarrow E_K(0^n)$

Let $M = M[1] \cdots M[m]$, where $|M[i]| = n$ for $1 \le i \le m-1$

for $i \leftarrow 1$ **to** $m-1$ **do**
$\quad X[i] \leftarrow M[i] \oplus \gamma_i \cdot L$
$\quad Y[i] \leftarrow E_K(X[i])$
$\Sigma \leftarrow Y[1] \oplus Y[2] \oplus \cdots \oplus Y[m-1] \oplus \mathsf{pad}(M[m])$
if $|M[m]| = n$ **then** $X[m] \leftarrow \Sigma \oplus L \cdot \mathsf{u}^{-1}$
$\qquad\qquad$ **else** $X[m] \leftarrow \Sigma$
$T \leftarrow E_K(X[m])$
return the leftmost τ bits of T

Fig. 4. PMAC-$\mathcal{T}_K(M)$

The PMAC$[E, \tau]$ tagging algorithm PMAC-$\mathcal{T}$ takes a k-bit key K and a message $M \in \{0,1\}^*$ as input and returns a τ-bit tag T. The algorithm of PMAC-$\mathcal{T}$ is described in Figure 4. See (1) for the definition of $\mathsf{pad}(\cdot)$, and Appendix for the computation of $\gamma_i \cdot L$.

The PMAC$[E, \tau]$ verification algorithm PMAC-$\mathcal{V}$ takes a k-bit key K, a message $M \in \{0,1\}^*$, and a tag $T \in \{0,1\}^\tau$ as input and returns a bit, and is defined in the natural way.

4   Side Channel Attacks and Their Countermeasures

In this section, we review side channel attacks and their countermeasures.

4.1   Side Channel Attacks

Side channel attacks (SCA) are a serious menace to embedded devices with cryptographic applications which are utilized in sensor and ad hoc networks. In SCA the attacker reveals secret information using side channel information such as power consumption while the victim device performs cryptographic applications [Koc96, KJJ99].

SCA includes two types of attacks; simple power analysis (SPA) and differential power analysis (DPA). In SPA the attacker utilizes power consumption directly, and parses it into a sequence composed of fundamental operations such as XOR and finite field operations. Then, he/she reveals the secret using the relation between the sequence and the secret. In DPA the attacker utilizes some statistical tools in addition to power consumption for revealing the secret. Normally he/she utilizes the average of power consumption for confirming his/her guess at the secret.

4.2   SCA on Block Ciphers

Kocher *et al.* were first to propose SCA on DES [Koc96, KJJ99]. Later SCA were extended to other block ciphers such as AES [Mes00a]. Whilst the several researches enhanced the attack, countermeasures against SCA were proposed. The masking method [Mes00a] is a typical example of a countermeasure on block ciphers.

Such countermeasures are not only theoretical or academic works, but also practical or industrial ones. In fact, nowadays some IC chips are equipped with tamper-resistant modules as countermeasures against SCA, especially countermeasures on a block cipher are utilized. MAC modules are achieved by using software combinations of such a block cipher with countermeasures. This is for reasons of the flexibility of cryptographic modules.

Whereas many researches have been made on SCA against block ciphers, little attention to SCA against MACs so far.

5 Proposed Attacks

In this section, we propose side channel attacks against MACs with tamper-resistant block cipher.

5.1 Simple Power Analysis

Simple Power Analysis on OMAC. We first present our SPA against OMAC.

In OMAC, the computation $L \cdot \mathtt{u}$ has a conditional branch as in (2), which is recognizable in view of SPA. If $\mathtt{msb}(L) = 0$ then only the left shift operation is performed. On the other hand, if $\mathtt{msb}(L) = 1$ then the left shift and XOR operations are performed. Thus, when an observed power consumption indicates XOR operation, the attacker can deduce the most significant bit of L is 1. That is, the attacker retrieves one-bit information on the secret L.

Figure 5 shows the experimental result for SPA on the computation $L \cdot \mathtt{u}$. For the experiment, the computation $L \cdot \mathtt{u}$ was implemented on an IC chip. While the computation $L \cdot \mathtt{u}$ was performed, the power consumption was observed by using an oscilloscope. Figure 5 shows this power consumption. The upper half is the power consumption when $\mathtt{msb}(L) = 0$, and the lower half is that when $\mathtt{msb}(L) = 1$. It is easy to recognize the additional operation in the lower half, which corresponds to XOR operation. Note that the power consumption curve just before XOR operation corresponds to the left shift operation $L \ll 1$, thus the counterpart appears in the upper half.

If the message length is not a positive multiple of n, then OMAC requires to compute $L \cdot \mathtt{u}^2$. In computing this value, we have only the left shift operation if $\mathtt{msb}(L \cdot \mathtt{u}) = 0$, while we have both the left shift and XOR operations if $\mathtt{msb}(L \cdot \mathtt{u}) = 1$. Therefore, the attacker can deduce the second most significant bit of L.

We have the following proposition.

Proposition 1. *There exists an SPA attacker A against OMAC that retrieves two-bit information on the secret L.*

Notice that L acts as a key (even if this value is derived from K). Namely, the security proof of OMAC requires L to be completely secret, and if L is retrieved by the attacker, then OMAC becomes insecure.

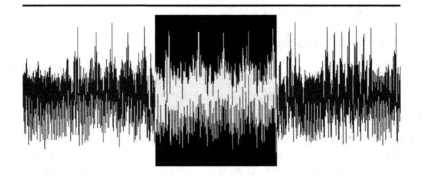

Fig. 5. SPA on $L \cdot u$

Simple Power Analysis on PMAC. We next present our SPA against PMAC. Now let $M = M[1] \circ M[2] \circ \cdots \circ M[m]$ be a message. To compute the tag for M, we have to compute the sequence of masks, $\gamma_1 \cdot L, \gamma_2 \cdot L, \ldots, \gamma_{m-1} \cdot L$. Now to compute $\gamma_i \cdot L$, we do as follows: $\gamma_i \cdot L = (\gamma_{i-1} \cdot L) \oplus (L \cdot u^{\mathsf{ntz}(i)})$, where $\mathsf{ntz}(i)$ is the number of trailing 0-bits in the binary representation of i (e.g., $\mathsf{ntz}(7) = 0$ and $\mathsf{ntz}(8) = 3$). The i-th word in the sequence $\gamma_1 \cdot L, \gamma_2 \cdot L, \gamma_3 \cdot L, \ldots$ is obtained by XORing the previous word with $(L \cdot u^{\mathsf{ntz}(i)})$ (See Appendix).

Therefore, to compute a tag for $M = M[1] \circ M[2] \circ \cdots \circ M[m]$, we have to compute $L \cdot u, L \cdot u^2, \ldots, L \cdot u^{\mathsf{ntz}(m-1)}$ and also $L \cdot u^{-1}$.

As we have seen in the case for OMAC, computing these values reveals $\mathsf{msb}(L), \mathsf{msb}(L \cdot u), \ldots, \mathsf{msb}(L \cdot u^{\mathsf{ntz}(m-1)-1})$ and also the least significant bit of L. Therefore, the attacker learns roughly $\log m$ bits of L. We have the following proposition.

Proposition 2. *There exists an SPA attacker A against PMAC that retrieves $\log m$-bit information on the secret L, where m demotes the block length of the message.*

5.2   Differential Power Analysis

First, we show the principle of the proposed DPA against MACs with tamper-resistant block cipher.

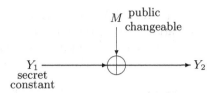

Fig. 6. DPA on XOR

Fig. 7. Experiment with DPA on XOR

Figure 6 shows the target XOR operation in a victim MAC. The XOR operation has two inputs Y_1 and M; Y_1 is secret and constant which the attacker tries to reveal, and M is public and the attacker can control this value. Y_2 is the output of the XOR, and the attacker does not know this value.

The first step of the attack is to guess a certain bit of Y_1, and he/she sorts the output Y_2 depending on the target bit of Y_2 is 0 or 1 according to the changeable input M. Then, the attacker observes the power consumption for the XOR operation with several inputs M. The third step is to compute the average power consumption for each group. Under the Hamming weight model [Mes00b], the power consumption depends on the Hamming weight of manipulated data. Hence the large power consumption implies that the target bit of Y_2 is 1 since the other bits behave as random and averaging eliminates the effect of the other bits. This provides the attacker with the information whether the original guess for the target bit of the secret Y_1 is correct or not. Repeating this procedure, the attacker can reveal the whole bits of the secret Y_1.

Note that once the attacker observes sufficient number of the power consumptions, he/she does not have to re-observe them for another target bit. Only he/she has to do is to re-classify Y_2 and compute the average power consumption for the new groups.

Figure 7 shows the experimental result for DPA on XOR. We utilized the same experimental environment as the one we did for SPA. For computing the averages, we observed the power consumption 100,000 times. Then, we classified the power consumption data into two groups, and computed the average power consumption for each group. Figure 7 shows the difference of these two averages. The peaks are easily recognized in the figure. Hence, when such an XOR operation exists in a victim MAC, the attack is realistic even if the underlying block cipher is tamper-resistant.

In what follows, we present our DPA against EMAC, OMAC, and PMAC. A remarkable aspect of our attack is that the attacker can achieve the selective forgery. That is, the attacker can choose the target message M^*. Furthermore, we allow the attacker to choose M^* *before* the key is chosen.

Differential Power Analysis on EMAC. The attack proceeds as follows.

1. A first chooses any massage M^*. Let $M^* = M^*[1] \circ M^*[2] \circ \cdots \circ M^*[m^*]$, where $|M^*[i]| = n$ for $1 \le i \le m^*$.
2. Then the oracle runs the key generation algorithm EMAC-$\mathcal{K}$ to choose a random secret key K. This K is hidden from A.
3. Now the attacker makes a tagging query $M_1 = M^*[1] \circ M^*[2] \circ \cdots \circ M^*[m^*-1]$ to EMAC-$\mathcal{T}_{K_1,K_2}(\cdot)$ oracle.
4. Then the attacker receives the associated side channel information. This side channel information includes the value of $Y[m^*-2]$. That is, the attacker uses $M^*[m^*-1]$ as the changeable input (M in Figure 6), and $Y[m^*-2]$ is treated as a secret constant (Y_1 in Figure 6).
5. A makes the next tagging query $M_2 = (Y[m^*-2] \oplus M^*[m^*-1]) \circ M^*[m^*]$ to get the tag $T_2 = $ EMAC-$\mathcal{T}_{K_1,K_2}(M_2)$.
6. Finally, A outputs (M^*, T^*), where $T^* \leftarrow T_2$, as a forgery attempt.

It is easy to verify that $\mathbf{Adv}^{\mathrm{mac}}_{\mathrm{EMAC}[E,\tau]}(A) = 1$ and A succeeds in selective forgery. Therefore, we have the following proposition.

Proposition 3. *There exists a DPA attacker A against EMAC such that* $\mathbf{Adv}^{\mathrm{mac}}\mathrm{EMAC}[E,\tau](A) = 1$.

Differential Power Analysis on OMAC. The attack proceeds similarly to the case for EMAC.

That is, A first chooses any massage $M^* = M^*[1] \circ M^*[2] \circ \cdots \circ M^*[m^*]$, where $|M^*[i]| = n$ for $1 \le i \le m^* - 1$. The length of $M^*[m^*]$ may be fewer than n bits. Then the oracle runs the key generation algorithm OMAC-$\mathcal{K}$ to choose a random secret key K. This K is hidden from A. Now the attacker makes a tagging query $M_1 = M^*[1] \circ M^*[2] \circ \cdots \circ M^*[m^*-1]$ to OMAC-$\mathcal{T}_K(\cdot)$ oracle.

Then the attacker learns the associated side channel information. As for the side channel information, there are two cases depending on the implementation of OMAC.

- $Y[m^*-2]$ is known to the attacker: In this case, the attacker uses $M^*[m^*-1]$ as the changeable input (M in Figure 6), and $Y[m^*-2]$ is treated as a secret constant (Y_1 in Figure 6). Then A proceeds as follows.
  1. A makes the second tagging query $M_2 = (Y[m^*-2] \oplus M^*[m^*-1]) \circ M^*[m^*]$ to get the second tag $T_2 = $ OMAC-$\mathcal{T}_K(M_2)$.
  2. Finally, A outputs (M^*, T^*), where $T^* \leftarrow T_2$, as a forgery attempt.
- $L \cdot \mathbf{u}$ is known to the attacker: In this case, the attacker uses $M^*[m^*-1]$ as the changeable input (M in Figure 6), and $L \cdot \mathbf{u}$ is treated as a secret constant (Y_1 in Figure 6). Then, A proceeds as follows.

1. A first computes $L = (L \cdot u) \cdot u^{-1}$. This can be done easily as shown in Sec. 2.
2. A makes the second tagging query $M_2 = 0^n \circ (M^*[1] \oplus L) \circ M^*[2] \circ M^*[3] \circ \cdots \circ M^*[m^*]$ to get the second tag $T_2 = \text{OMAC-}\mathcal{T}_K(M_2)$.
3. Finally, A outputs (M^*, T^*), where $T^* \leftarrow T_2$, as a forgery attempt.

It is easy to verify that $\mathbf{Adv}^{\text{mac}}_{\text{OMAC}[E,\tau]}(A) = 1$ and A succeeds in selective forgery. We have the following proposition.

Proposition 4. *There exists a DPA attacker A against OMAC such that* $\mathbf{Adv}^{\text{mac}}_{\text{OMAC}[E,\tau]}(A) = 1$.

Differential Power Analysis on PMAC. In this section, we present our DPA against PMAC. Let $M^* = M^*[1] \circ M^*[2] \circ \cdots \circ M^*[m^*]$ be a message that the attacker wants to forge, where $|M^*[i]| = n$ for $1 \leq i \leq m^* - 1$. The length of $M^*[m^*]$ may be fewer than n bits.

The attack proceeds as follows.

1. The attacker makes a tagging query $M_1 = M^*[1] \circ M^*[2] \circ \cdots \circ M^*[m^* - 1] \circ M$ to PMAC-$\mathcal{T}_K(\cdot)$ oracle. If $|M^*[m^*]| = n$, then we choose $M \neq M^*[m^*]$ and $|M| = n$. Otherwise we choose $M \neq M^*[m^*]$ and $1 \leq |M| < n$.
2. Then the attacker receives the associated side channel information Σ_1.
3. A makes the second tagging query $M_2 = M'[1] \circ M'[2] \circ \cdots \circ M'[m^* - 1] \circ M'[m^*]$ to PMAC-$\mathcal{T}_K(\cdot)$ oracle, where $M_2 \neq M_1$ and $M_2 \neq M^*$. If $|M^*[m^*]| = n$, then we choose $M'[m^*] \neq M^*[m^*]$ and $|M'[m^*]| = n$. Otherwise we choose $M'[m^*] \neq M^*[m^*]$ and $1 \leq |M'[m^*]| < n$.
4. Then the attacker receives the associated side channel information Σ_2.
5. Let M'' be a string such that $\mathsf{pad}(M'') = \Sigma_2 \oplus \Sigma_1 \oplus \mathsf{pad}(M^*[m^*])$. Then A makes the third tagging query $M_3 = M'[1] \circ M'[2] \circ \cdots \circ M'[m^* - 1] \circ M''$ to PMAC-$\mathcal{T}_K(\cdot)$ oracle to receive T_3.
6. Finally, A outputs (M^*, T^*), where $T^* \leftarrow T_3$, as a forgery attempt.

If $\Sigma_2 \oplus \Sigma_1 \oplus \mathsf{pad}(M^*[m^*]) = 0^n$, then the attack fails since there is no M''. But this occurs with probability $1/2^n$, assuming that the underlying block cipher behaves as a random permutation. Therefore, we have the following proposition.

Proposition 5. *There exists a DPA attacker A against PMAC such that* $\mathbf{Adv}^{\text{mac}}_{\text{PMAC}[E,\tau]}(A) \geq 1 - 1/2^n$.

6   Conclusion

In this paper we showed that side channel attacks can be applied to MACs, even if the countermeasure is applied to the underlying block cipher. We showed that EMAC, OMAC, and PMAC are vulnerable to our attacks. We showed that, based on SPA, information on the keys can be extracted. Also, based on DPA, we presented selective forgery against these MACs, assuming that the attacker is in a chosen plaintext attack scenario. Our results suggest that protecting block ciphers against side channel attacks is not sufficient, and the countermeasure is needed for modes implementations.

References

[BKR00]   M. Bellare, J. Kilian, and P. Rogaway, "The Security of the Cipher Block Chaining Message Authentication Code," *JCSS,* Vol. 61, No. 3, pp. 362–399, 2000. Earlier version in CRYPTO '94, LNCS 839, pp. 341–358, 1994.

[BBB+95]   A. Berendschot, B. den Boer, J. P. Boly, A. Bosselaers, J. Brandt, D. Chaum, I. Damgård, M. Dichtl, W. Fumy, M. van der Ham, C. J. A. Jansen, P. Landrock, B. Preneel, G. Roelofsen, P. de Rooij, and J. Vandewalle. "Final Report of RACE Integrity Primitives. " LNCS 1007, 1995.

[BR02]   J. Black and P. Rogaway, "A Block-Cipher Mode of Operation for Parallelizable Message Authentication," EUROCRYPT 2002, LNCS 2332, pp. 384–397.

[BU02]   J. Black, and H. Urtubia, "Side-Channel Attacks on Symmetric Encryption Schemes: The Case for Authenticated Encryption," In Proc. of 11th USENIX security symposium, pp. 327–338, 2002.

[CHVV03]   B. Canvel, A. Hiltgen, S. Vaudenay, and M. Vaugnoux, "Password Interception in a SSL/TLS Channel," CRYPTO 2003, LNCS 2729, pp. 583–599, 2003.

[DR02]   J. Daemen and V. Rijmen, "The Design of Rijndael," Springer-Verlag, Berlin Germany, 2002.

[FIPS94]   FIPS 113, Computer data authentication. Federal Information Processing Standards Publication 113, U.S. Department of Commerce / National Bureau of Standards, National Technical Information Service, Springfield, Virginia, 1994.

[IEEE]   IEEE 802.16 Task Group e (Mobile WirelessMAN), http://wirelessman.org/tge/.

[ISO99]   ISO/IEC 9797-1, Information technology — security techniques — data integrity mechanism using a cryptographic check function employing a block cipher algorithm. Organization for Standards, Geneva, Switzerland, 1999. Second edition.

[IK03]   T. Iwata and K. Kurosawa, "OMAC: One-Key CBC MAC," FSE 2003, LNCS 2887, pp. 129–153, 2003.

[KR03]   V. Klima and T. Rosa, "Side Channel Attacks on CBC Encrypted Messages in the PKCS#7 Format," IACR ePrint Archive 2003/098, 2003.

[Koc96]   C. Kocher, "Timing attacks on Implementations of Diffie-Hellman, RSA, DSS, and other Systems," CRYPTO '96, LNCS 1109, pp. 104–113, 1996.

[KJJ99]   C. Kocher, J. Jaffe, and B. Jun, "Differential Power Analysis," CRYPTO '99, LNCS 1666, pp. 388–397, 1999.

[Mes00a]   T. Messerges, "Securing the AES Finalists against Power Analysis Attacks," FSE 2000, LNCS 1978, pp. 150–164, 2001.

[Mes00b]   T. Messerges, "Using Second-Order Power Analysis to Attack DPA Resistant Software," CHES 2000, LNCS 1965, pp. 238–251, 2000.

[MDS02]   T.S. Messerges, E.A. Dabbish, R.H. Sloan, "Examining Smart-Card Security under the Threat of Power Analysis Attacks," IEEE Trans. Computers, Vol. 51, No. 5, pp. 541–552, 2002.

[Möl04]   B. Möller, "Security of CBC Ciphersuites in SSL/TLS: Problems and Countermeasures," Available at http://www.openssl.org/~bodo/tls-cbc.txt, 2004.

[NIST]   M. Dworkin, "Recommendation for Block Cipher Modes of Operation: The CMAC Mode for Authentication," Available at http://csrc.nist.gov/publications/nistpubs/800-38B/SP_800-38B.pdf, 2005.

[PY04]   K.G. Paterson, and A. Yau, "Padding Oracle Attacks on the ISO CBC Mode Encryption Standard," CT-RSA 2004, LNCS 2964, pp. 305–323, 2004.

[Vau02]   S. Vaudenay, "Security Flaws Induced by CBC Padding - Applications to SSL, IPSEC, WTLS," EUROCRYPT 2002, LNCS 2332, pp. 534–545, 2002.

A   Gray Code

For any $l \geq 1$, a Gray code is an ordering $\gamma^l = \gamma_0^l \gamma_1^l \cdots \gamma_{2^l-1}^l$ of $\{0,1\}^l$ such that successive points differ (in the Hamming sense) by just one bit. For n a fixed number, PMAC uses the "canonical" Gray code $\gamma = \gamma^n$ constructed by $\gamma^1 = 0 \circ 1$ while, for $l > 0$,

$$\gamma^{l+1} = (0\gamma_0^l) \circ (0\gamma_1^l) \circ \cdots \circ (0\gamma_{2^l-2}^l) \circ (0\gamma_{2^l-1}^l) \circ (1\gamma_{2^l-1}^l) \circ (1\gamma_{2^l-2}^l) \circ \cdots \circ (1\gamma_1^l) \circ (1\gamma_1^l) .$$

It is easy to see that γ is a Gray code. What is more, for $1 \leq i \leq 2^n - 1$, $\gamma_i = \gamma_{i-1} \oplus (0^{n-1}1 \ll \mathsf{ntz}(i))$, where $\mathsf{ntz}(i)$ is the number of trailing 0-bits in the binary representation of i (e.g., $\mathsf{ntz}(7) = 0$ and $\mathsf{ntz}(8) = 3$). This makes easy to compute successive points.

Let $L \in \{0,1\}^n$ and consider the problem of successively forming the strings $\gamma_1 \cdot L, \gamma_2 \cdot L, \ldots, \gamma_m \cdot L$. Of course $\gamma_1 \cdot L = 1 \cdot L = L$. Now for $i \geq 2$, assume one has already produced $\gamma_{i-1} \cdot L$. Since $\gamma_i = \gamma_{i-1} \oplus (0^{n-1}1 \ll \mathsf{ntz}(i))$, we know that $\gamma_i \cdot L = (\gamma_{i-1} \oplus (0^{n-1}1 \ll \mathsf{ntz}(i))) \cdot L = (\gamma_{i-1} \cdot L) \oplus (0^{n-1}1 \ll \mathsf{ntz}(i)) \cdot L = (\gamma_{i-1} \cdot L) \oplus (L \cdot \mathsf{u}^{\mathsf{ntz}(i)})$. That is, the i-th word in the sequence $\gamma_1 \cdot L, \gamma_2 \cdot L, \gamma_3 \cdot L, \ldots$ is obtained by XORing the previous word with $(L \cdot \mathsf{u}^{\mathsf{ntz}(i)})$.

Author Index

Lecture Notes in Computer Science

For information about Vols. 1–3742

please contact your bookseller or Springer

Vol. 3791: A. Adi, S. Stoutenburg, S. Tabet (Eds.), Rules and Rule Markup Languages for the Semantic Web. X, 225 pages. 2005.

Vol. 3790: G. Alonso (Ed.), Middleware 2005. XIII, 443 pages. 2005.

Vol. 3789: A. Gelbukh, Á. de Albornoz, H. Terashima-Marín (Eds.), MICAI 2005: Advances in Artificial Intelligence. XXVI, 1198 pages. 2005. (Sublibrary LNAI).

Vol. 3788: B. Roy (Ed.), Advances in Cryptology - ASIACRYPT 2005. XIV, 703 pages. 2005.

Vol. 3785: K.-K. Lau, R. Banach (Eds.), Formal Methods and Software Engineering. XIV, 496 pages. 2005.

Vol. 3784: J. Tao, T. Tan, R.W. Picard (Eds.), Affective Computing and Intelligent Interaction. XIX, 1008 pages. 2005.

Vol. 3783: S. Qing, W. Mao, J. Lopez, G. Wang (Eds.), Information and Communications Security. XIV, 492 pages. 2005.

Vol. 3781: S.Z. Li, Z. Sun, T. Tan, S. Pankanti, G. Chollet, D. Zhang (Eds.), Advances in Biometric Person Authentication. XI, 250 pages. 2005.

Vol. 3780: K. Yi (Ed.), Programming Languages and Systems. XI, 435 pages. 2005.

Vol. 3779: H. Jin, D. Reed, W. Jiang (Eds.), Network and Parallel Computing. XV, 513 pages. 2005.

Vol. 3778: C. Atkinson, C. Bunse, H.-G. Gross, C. Peper (Eds.), Component-Based Software Development for Embedded Systems. VIII, 345 pages. 2005.

Vol. 3777: O.B. Lupanov, O.M. Kasim-Zade, A.V. Chaskin, K. Steinhöfel (Eds.), Stochastic Algorithms: Foundations and Applications. VIII, 239 pages. 2005.

Vol. 3776: S.K. Pal, S. Bandyopadhyay, S. Biswas (Eds.), Pattern Recognition and Machine Intelligence. XXIV, 808 pages. 2005.

Vol. 3775: J. Schönwälder, J. Serrat (Eds.), Ambient Networks. XIII, 281 pages. 2005.

Vol. 3774: G. Bierman, C. Koch (Eds.), Database Programming Languages. X, 295 pages. 2005.

Vol. 3773: A. Sanfeliu, M.L. Cortés (Eds.), Progress in Pattern Recognition, Image Analysis and Applications. XX, 1094 pages. 2005.

Vol. 3772: M. Consens, G. Navarro (Eds.), String Processing and Information Retrieval. XIV, 406 pages. 2005.

Vol. 3771: J.M.T. Romijn, G.P. Smith, J. van de Pol (Eds.), Integrated Formal Methods. XI, 407 pages. 2005.

Vol. 3770: J. Akoka, S.W. Liddle, I.-Y. Song, M. Bertolotto, I. Comyn-Wattiau, W.-J. van den Heuvel, M. Kolp, J. Trujillo, C. Kop, H.C. Mayr (Eds.), Perspectives in Conceptual Modeling. XXII, 476 pages. 2005.

Vol. 3769: D.A. Bader, M. Parashar, V. Sridhar, V.K. Prasanna (Eds.), High Performance Computing – HiPC 2005. XXVIII, 550 pages. 2005.

Vol. 3768: Y.-S. Ho, H.J. Kim (Eds.), Advances in Multimedia Information Processing - PCM 2005, Part II. XXVIII, 1088 pages. 2005.

Vol. 3767: Y.-S. Ho, H.J. Kim (Eds.), Advances in Multimedia Information Processing - PCM 2005, Part I. XXVIII, 1022 pages. 2005.

Vol. 3766: N. Sebe, M.S. Lew, T.S. Huang (Eds.), Computer Vision in Human-Computer Interaction. X, 231 pages. 2005.

Vol. 3765: Y. Liu, T. Jiang, C. Zhang (Eds.), Computer Vision for Biomedical Image Applications. X, 563 pages. 2005.

Vol. 3764: S. Tixeuil, T. Herman (Eds.), Self-Stabilizing Systems. VIII, 229 pages. 2005.

Vol. 3762: R. Meersman, Z. Tari, P. Herrero (Eds.), On the Move to Meaningful Internet Systems 2005: OTM 2005 Workshops. XXXI, 1228 pages. 2005.

Vol. 3761: R. Meersman, Z. Tari (Eds.), On the Move to Meaningful Internet Systems 2005: CoopIS, DOA, and ODBASE, Part II. XXVII, 653 pages. 2005.

Vol. 3760: R. Meersman, Z. Tari (Eds.), On the Move to Meaningful Internet Systems 2005: CoopIS, DOA, and ODBASE, Part I. XXVII, 921 pages. 2005.

Vol. 3759: G. Chen, Y. Pan, M. Guo, J. Lu (Eds.), Parallel and Distributed Processing and Applications - ISPA 2005 Workshops. XIII, 669 pages. 2005.

Vol. 3758: Y. Pan, D.-x. Chen, M. Guo, J. Cao, J.J. Dongarra (Eds.), Parallel and Distributed Processing and Applications. XXIII, 1162 pages. 2005.

Vol. 3757: A. Rangarajan, B. Vemuri, A.L. Yuille (Eds.), Energy Minimization Methods in Computer Vision and Pattern Recognition. XII, 666 pages. 2005.

Vol. 3756: J. Cao, W. Nejdl, M. Xu (Eds.), Advanced Parallel Processing Technologies. XIV, 526 pages. 2005.

Vol. 3754: J. Dalmau Royo, G. Hasegawa (Eds.), Management of Multimedia Networks and Services. XII, 384 pages. 2005.

Vol. 3753: O.F. Olsen, L.M.J. Florack, A. Kuijper (Eds.), Deep Structure, Singularities, and Computer Vision. X, 259 pages. 2005.

Vol. 3752: N. Paragios, O. Faugeras, T. Chan, C. Schnörr (Eds.), Variational, Geometric, and Level Set Methods in Computer Vision. XI, 369 pages. 2005.

Vol. 3751: T. Magedanz, E.R.M. Madeira, P. Dini (Eds.), Operations and Management in IP-Based Networks. X, 213 pages. 2005.

Vol. 3750: J.S. Duncan, G. Gerig (Eds.), Medical Image Computing and Computer-Assisted Intervention – MICCAI 2005, Part II. XL, 1018 pages. 2005.

Vol. 3749: J.S. Duncan, G. Gerig (Eds.), Medical Image Computing and Computer-Assisted Intervention – MICCAI 2005, Part I. XXXIX, 942 pages. 2005.

Vol. 3748: A. Hartman, D. Kreische (Eds.), Model Driven Architecture – Foundations and Applications. IX, 349 pages. 2005.

Vol. 3747: C.A. Maziero, J.G. Silva, A.M.S. Andrade, F.M.d. Assis Silva (Eds.), Dependable Computing. XV, 267 pages. 2005.

Vol. 3746: P. Bozanis, E.N. Houstis (Eds.), Advances in Informatics. XIX, 879 pages. 2005.

Vol. 3745: J.L. Oliveira, V. Maojo, F. Martín-Sánchez, A.S. Pereira (Eds.), Biological and Medical Data Analysis. XII, 422 pages. 2005. (Sublibrary LNBI).

Vol. 3744: T. Magedanz, A. Karmouch, S. Pierre, I.S. Venieris (Eds.), Mobility Aware Technologies and Applications. XIV, 418 pages. 2005.